This book was made
possible by a donation from
Mr. Jim McUsic, winner of
the
2011 Marcus R. Tower
Service Award.

tulsa
LIBRARY TRUST

FRONTIERS OF SCIENCE

COMPUTER SCIENCE

Notable Research and Discoveries

KYLE KIRKLAND, PH.D.

Facts On File
An imprint of Infobase Publishing

COMPUTER SCIENCE: Notable Research and Discoveries

Facts On File, Inc.
An imprint of Infobase Publishing
132 West 31st Street
New York NY 10001

Library of Congress Cataloging-in-Publication Data

Kirkland, Kyle.
    Computer science: notable research and discoveries / Kyle Kirkland.
        p. cm.—(Frontiers of science)
    Includes bibliographical references and index.
    ISBN 978-0-8160-7441-9
    1. Computer science—Research. 2. Computers—Research. I. Title.
    QA76.27.K58 2010
    004.072—dc22   2009025411

Facts On File books are available at special discounts when purchased in bulk quantities for businesses, associations, institutions, or sales promotions. Please call our Special Sales Department in New York at (212) 967-8800 or (800) 322-8755.

You can find Facts On File on the World Wide Web at http://www.factsonfile.com

Text design by Kerry Casey
Illustrations by Sholto Ainslie
Photo research by Tobi Zausner, Ph.D.
Composition by Mary Susan Ryan-Flynn
Cover printed by Bang Printing, Inc., Brainerd, Minn.
Book printed and bound by Bang Printing, Inc., Brainerd, Minn.
Date printed: May 2010

Printed in the United States of America

10 9 8 7 6 5 4 3 2 1

This book is printed on acid-free paper.

CONTENTS

PREFACE

Discovering what lies behind a hill or beyond a neighborhood can be as simple as taking a short walk. But curiosity and the urge to make new discoveries usually require people to undertake journeys much more adventuresome than a short walk, and scientists often study realms far removed from everyday observation—sometimes even beyond the present means of travel or vision. Polish astronomer Nicolaus Copernicus's (1473–1543) heliocentric (Sun-centered) model of the solar system, published in 1543, ushered in the modern age of astronomy more than 400 years before the first rocket escaped Earth's gravity. Scientists today probe the tiny domain of atoms, pilot submersibles into marine trenches far beneath the waves, and analyze processes occurring deep within stars.

Many of the newest areas of scientific research involve objects or places that are not easily accessible, if at all. These objects may be trillions of miles away, such as the newly discovered planetary systems, or they may be as close as inside a person's head; the brain, a delicate organ encased and protected by the skull, has frustrated many of the best efforts of biologists until recently. The subject of interest may not be at a vast distance or concealed by a protective covering, but instead it may be removed in terms of time. For example, people need to learn about the evolution of Earth's weather and climate in order to understand the changes taking place today, yet no one can revisit the past.

Frontiers of Science is an eight-volume set that explores topics at the forefront of research in the following sciences:

- biological sciences
- chemistry
- computer science

- Earth science
- marine science
- physics
- space and astronomy
- weather and climate

The set focuses on the methods and imagination of people who are pushing the boundaries of science by investigating subjects that are not readily observable or are otherwise cloaked in mystery. Each volume includes six topics, one per chapter, and each chapter has the same format and structure. The chapter provides a chronology of the topic and establishes its scientific and social relevance, discusses the critical questions and the research techniques designed to answer these questions, describes what scientists have learned and may learn in the future, highlights the technological applications of this knowledge, and makes recommendations for further reading. The topics cover a broad spectrum of the science, from issues that are making headlines to ones that are not as yet well known. Each chapter can be read independently; some overlap among chapters of the same volume is unavoidable, so a small amount of repetition is necessary for each chapter to stand alone. But the repetition is minimal, and cross-references are used as appropriate.

Scientific inquiry demands a number of skills. The National Committee on Science Education Standards and Assessment and the National Research Council, in addition to other organizations such as the National Science Teachers Association, have stressed the training and development of these skills. Science students must learn how to raise important questions, design the tools or experiments necessary to answer these questions, apply models in explaining the results and revise the model as needed, be alert to alternative explanations, and construct and analyze arguments for and against competing models.

Progress in science often involves deciding which competing theory, model, or viewpoint provides the best explanation. For example, a major issue in biology for many decades was determining if the brain functions as a whole (the holistic model) or if parts of the brain carry out specialized functions (functional localization). Recent developments in brain imaging resolved part of this issue in favor of functional localization by showing that specific regions of the brain are more active during certain tasks. At the same time, however, these experiments have raised other questions that future research must answer.

The logic and precision of science are elegant, but applying scientific skills can be daunting at first. The goals of the Frontiers of Science set are to explain how scientists tackle difficult research issues and to describe recent advances made in these fields. Understanding the science behind the advances is critical because sometimes new knowledge and theories seem unbelievable until the underlying methods become clear. Consider the following examples. Some scientists have claimed that the last few years are the warmest in the past 500 or even 1,000 years, but reliable temperature records date only from about 1850. Geologists talk of volcano hot spots and plumes of abnormally hot rock rising through deep channels, although no one has drilled more than a few miles below the surface. Teams of neuroscientists—scientists who study the brain—display images of the activity of the brain as a person dreams, yet the subject's skull has not been breached. Scientists often debate the validity of new experiments and theories, and a proper evaluation requires an understanding of the reasoning and technology that support or refute the arguments.

Curiosity about how scientists came to know what they do—and why they are convinced that their beliefs are true—has always motivated me to study not just the facts and theories but also the reasons why these are true (or at least believed). I could never accept unsupported statements or confine my attention to one scientific discipline. When I was young, I learned many things from my father, a physicist who specialized in engineering mechanics, and my mother, a mathematician and computer systems analyst. And from an archaeologist who lived down the street, I learned one of the reasons why people believe Earth has evolved and changed—he took me to a field where we found marine fossils such as shark's teeth, which backed his claim that this area had once been under water! After studying electronics while I was in the air force, I attended college, switching my major a number of times until becoming captivated with a subject that was itself a melding of two disciplines—biological psychology. I went on to earn a doctorate in neuroscience, studying under physicists, computer scientists, chemists, anatomists, geneticists, physiologists, and mathematicians. My broad interests and background have served me well as a science writer, giving me the confidence, or perhaps I should say chutzpah, to write a set of books on such a vast array of topics.

Seekers of knowledge satisfy their curiosity about how the world and its organisms work, but the applications of science are not limited to intellectual achievement. The topics in Frontiers of Science affect society on a multitude of levels. Civilization has always faced an uphill battle to procure scarce resources, solve technical problems, and maintain order. In modern times, one of the most important resources is energy, and the physics of fusion potentially offers a nearly boundless supply. Technology makes life easier and solves many of today's problems, and nanotechnology may extend the range of devices into extremely small sizes. Protecting one's personal information in transactions conducted via the Internet is a crucial application of computer science.

But the scope of science today is so vast that no set of eight volumes can hope to cover all of the frontiers. The chapters in Frontiers of Science span a broad range of each science but could not possibly be exhaustive. Selectivity was painful (and editorially enforced) but necessary, and in my opinion, the choices are diverse and reflect current trends. The same is true for the subjects within each chapter—a lot of fascinating research did not get mentioned, not because it is unimportant, but because there was no room to do it justice.

Extending the limits of knowledge relies on basic science skills as well as ingenuity in asking and answering the right questions. The 48 topics discussed in these books are not straightforward laboratory exercises but complex, gritty research problems at the frontiers of science. Exploring uncharted territory presents exceptional challenges but also offers equally impressive rewards, whether the motivation is to solve a practical problem or to gain a better understanding of human nature. If this set encourages some of its readers to plunge into a scientific frontier and conquer a few of its unknowns, the books will be worth all the effort required to produce them.

ACKNOWLEDGMENTS

Thanks go to Frank K. Darmstadt, executive editor at Facts On File, and the rest of the staff for all their hard work, which I admit I sometimes made a little bit harder. Thanks also to Tobi Zausner for researching and locating so many great photographs. I also appreciate the time and effort of a large number of researchers who were kind enough to pass along a research paper or help me track down some information.

INTRODUCTION

Computer science is a rapidly growing discipline that continues to awe, inspire, and intimidate all those who interact with its many elements and capabilities. In the 19th century, the term *computer* referred to people who performed mathematical computations. But mechanical tabulating machines and calculators began to appear in the late 19th century and early 20th century, and in 1946, engineers J. Presper Eckert (1919–95) and John Mauchly (1907–80) built one of the first modern electronic computers, known as the Electronic Numerical Integrator and Computer (ENIAC). ENIAC was an important advance but had some disadvantages—it was the size of a room, ran slowly, and often suffered failures in its electrical components. But since the 1940s, computers have evolved into fast and efficient machines that fill almost every niche in today's society. Computers are used in businesses and at home, underlie extensive communication networks such as the Internet, help pilots fly airplanes and astronauts fly the space shuttle, maneuver a vehicle on Mars, sort mail based on typed or handwritten addresses, and much else.

The expanding role of computers has begun to encroach on tasks that require substantial thought—at least for a person. For example, in 1997, a computer called Deep Blue defeated Garry Kasparov, the reigning World Chess Champion at the time, in a chess match. Chess-playing computer programs have been routinely defeating novice chess players since the 1970s, but Deep Blue beat one of the best.

No one is certain how much more powerful—and possibly intelligent—computers will become in the 21st century. *Computer Science,* one volume of the multivolume Frontiers of Science set, explores six prominent topics in computer science research that address issues

This construction worker is making notes with a portable computer. *(Justin Horrocks/iStockphoto)*

concerning the capacity of computers and their applications and describes how computer scientists conduct research and attempt to formulate answers to important questions. Research published in journals, presented at conferences, and described in news releases supply plenty of examples of scientific problems and how researchers attempt to solve them. Although these reports are summarized and presented briefly in this book, they offer the student and the general reader insight into the methods and applications of computer science research.

Students need to keep up with the latest developments in these quickly moving fields of research, but they have difficulty finding a source that explains the basic concepts while discussing the background and context essential for the "big picture." This book describes the evolution of each of the six main topics it covers, and explains the problems that researchers are currently investigating as well as the methods they are developing to solve them.

Although a computer may perform intelligent tasks, the performance of most machines today reflects the skill of computer engineers and programmers. None of the applications mentioned above would have been possible without the efforts of computer engineers who design the machines, and computer programmers who write the programs to provide the necessary instructions. Most computers today perform a series of simple steps, and must be told exactly which steps to perform and in what order. Deep Blue, for example, did not think as a person does, but instead ran a program to search for the best move, as determined by complicated formulas. A fast computer such as Deep Blue can zip through these instructions so quickly that it is capable of impressive feats of "intelligence." Chapter 1 describes how computer scientists have built and designed these machines, and what else may be possible in the future.

But some computer scientists are working on making computers smarter—and more like humans. The human brain consists of complex *neural networks* that process sensory information, extract important features, and solve problems. Chapter 2 discusses research aimed at mimicking neural networks with computer technology.

Speedy computations are essential in many of these operations, and fast computers can find solutions to complicated problems. Deep Blue's program, for instance, churned through millions of instructions every second to find the optimal chess move. But certain kinds of problems have remained intractable, even with the fastest computers. Many of these problems, such as factoring integers or finding the shortest distances in certain routes, have important practical applications for engineering and science, as well as for computer networks and economics. People can program computers to address these problems on a small scale—factoring a small number such as 20, or finding a route with only three cities to visit—but problems involving larger numbers require too much time. A fundamental question in computer science is whether efficient solutions generally exist for these problems, as discussed in chapter 3.

An efficient method to solve these problems, if one is ever found, would have a tremendous impact, especially on the Internet. Personal and confidential information, such as credit card numbers, gets passed from computer to computer every day on the Internet. This information must be protected by making the information unreadable to all except the intended recipient. The science of writing and reading secret messages is called *cryptology,* and many techniques today could be broken—and their secrets exposed—if an efficient solution is found to the problems described in chapter 3. Cryptology is the subject of chapter 4.

One of the most important human senses is vision. Images provide a wealth of information that is difficult or cumbersome to put into words. These days, images are often processed in digital form—arrays of numbers that computers can store and process. As computers become faster and smarter, people have started using these machines to perform functions similar to human vision, such as reading. These projects are the central theme of chapter 5.

Searching for patterns is an integral part of many computer applications—for example, looking for clues to crack a secret message, or sifting through the features of an image to find a specific object. Biologists have recently amassed a huge quantity of data involving genetics. Patterns in this kind of information contain vital clues about how organisms develop, what traits they have, and how certain diseases arise and progress. Overwhelmed by the sheer size of these data, which is the equivalent of thousands of encyclopedia volumes, biologists have turned to computer science for help. Chapter 6 examines the results so far and the ongoing research efforts.

Although these six chapters do not cover all of the frontiers of computer science, they offer a broad spectrum of significant topics, each of which has the potential to change how computers are used—and how these machines will continue transforming the way people live and work. Computers have made life easier in many ways, relieving people of boring and time-consuming tasks, but computers have also made life more complicated, forcing people to keep up with technological developments. This book will help students and other readers understand how computer scientists conduct research, and where this research is heading.

Advanced Computers

A fundamental element of research in computer science is the computer itself. Despite the efficiency of today's machines, the computer remains a frontier of science. The reason for this is the same as it was during the early years of computational technology.

In 1790, marshals of the newly formed government of the United States set out on horseback to perform the important mission of counting the country's population. Taking an accurate census was essential in order to apportion the number of congressional delegates for each district, as specified by the U.S. Constitution. According to the U.S. Census Bureau, the census-takers manually compiled a list of 3,929,214 people in less than a year. Officials took another census each decade, and by 1880 the population had grown to 50,155,783. But census-takers had reached the breaking point—it took them almost the whole decade to finish tabulating the 1880 census, and the country continued to grow at an astonishing rate. Government officials feared that the 1890 census would not be completed before they had to begin the 1900 census.

The solution to this problem was automation. In response to a competition sponsored by the Bureau of the Census, Herman Hollerith (1860–1929), a young engineer, designed an automatic "census counting machine." Census personnel collected *data*—the plural of a Latin word, *datum,* meaning information—and encoded the information in the positions of holes punched in cards. These cards were the same size as dollar bills of the time, meaning that a stack of cards conveniently fit into boxes used by the Treasury Department. When operators inserted the cards into the machine, an electromechanical process automatically tabulated the population figures. Using Hollerith's machines, the 1890 census of 62,979,766 people was counted within a few months, and the statistical tables were completed two years later.

Hollerith formed a company, the Tabulating Machine Company, in 1896. The company changed its name in 1924 to International Business Machines (IBM) Corporation. IBM thrived, and is presently one of the world's largest companies.

Computational machines have also thrived. The need for speed and efficiency—the same needs of the 1890 census—motivated the development of computers into the ubiquitous machines they are today. Computers are in homes, offices, cars, and even spacecraft, and people carry portable computers known as notebooks or laptops whenever they travel. Yet the evolution of computers is by no means finished. One of the most active frontiers of computer science is the development of faster and more efficient computers, which may eventually transform the world as drastically as their predecessors did. This chapter describes research at the forefront of computer systems, including optical and molecular computers and computers based on sophisticated concepts in physics.

INTRODUCTION

The idea of using machines for calculation did not begin with Hollerith. In addition to using fingers, toes, and little marks on clay tablets or parchment to help make arithmetic easier, people in early times used a tool called an *abacus*. This device consisted of a system of beads sliding along wires or strings attached to a frame. An abacus did not perform the calculations, but helped people keep track of the totals with the movable beads. Babylonians, Greeks, and Romans, among other ancient peoples, developed and used abaci (plural of abacus) several thousand years ago.

French scientist and inventor Blaise Pascal (1623–62) invented one of the earliest adding machines in 1642. Pascal's motivation to build this device would be familiar to citizens today—tax calculations. His machine consisted of a series of interconnected wheels, with numbers etched on the rims. The gears connecting the wheels were designed to advance when the adjacent wheel made a complete revolution, similar to the operation of the odometer (mileage indicator) of a car. Although successful, Pascal's machine was too expensive for everyday use.

British mathematician Charles Babbage (1791–1871) designed devices in the 19th century that were the forerunners of the modern computer—and would have been the earliest computers, had Babbage been able to obtain funds to build them. His first effort, called a "difference engine," was to be a calculating machine. (The term *difference* in the name was due to a numerical technique called the differences method that the machine would employ.) The machine was well designed but complicated, and manufacturing difficulties exhausted the money Babbage had acquired before construction was complete.

Babbage's next design was for an even more ambitious machine that would have served a variety of purposes instead of fulfilling a single function (such as calculating). In his book *Passages from the Life of a Philosopher,* published in 1864, Babbage wrote, "The whole of arithmetic now appeared within the grasp of the mechanism." This machine, called an "analytical engine," was to be programmable—information stored on punched cards would direct the machine's operation, so that it could be programmed to solve different problems. Although Babbage also failed to finish constructing the analytical engine, the idea of an efficient, general-purpose machine presaged the computers of today. As Babbage wrote in 1864, "As soon as an Analytical Engine exists, it will necessarily guide the future course of the science."

In Babbage's time, and on into the early 20th century, the term *computer* referred to a person who performed mathematical calculations. For example, Harvard Observatory employed "computers"— often women—who compiled catalogs of stars based on the observatory's astronomical data. (Some of these "computers," such as Annie Jump Cannon, went on to become astronomers.) Hollerith's machine, described above, was a highly useful calculating machine as well as an important advance in computational technology, but it was not a versatile, programmable device.

Harvard University was the site of one of the earliest machines that could be called a computer in the modern meaning of the word. Guided by engineer Howard Aiken (1900–73), IBM manufactured the components of the Automatic Sequence Controlled Calculator, also known as the Mark I, which engineers assembled at Harvard University in 1944. Using both electrical and mechanical parts, the Mark I was 51 feet (15.5 m) in length, eight feet (2.4 m) in height, and weighed a whopping 10,000 pounds (22,000 kg). A year later, the University of

Pennsylvania finished a completely electronic computer, known as the Electronic Numerical Integrator and Computer (ENIAC). Designed by engineers John Mauchly (1907–80) and J. Presper Eckert (1919–95), ENIAC needed more than 1,000 square feet (93 m^2) of space, and plenty of fans to keep the components from overheating. British researchers had built similar machines, known as Colossus computers, a little earlier, but the government kept their operation secret because they were used to read the enemy's secret messages during World War II.

These large computers, known as *mainframes,* received their programming instructions from punched cards or tape. Computations such as ballistic tables—the calculation of artillery trajectories based on wind, distance, and other factors, as needed by the U.S. military during World War II—could be accomplished in a fraction of the time required for manual tabulation. ENIAC, for instance, was capable of

ENIAC, attended by several technicians and its designers, J. Presper Eckert (foreground left) and John Mauchly (leaning against the column) *(AP Photo/ University of PA)*

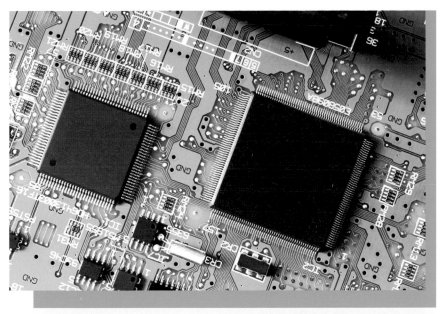

Integrated circuits mounted on a green circuit board *(Roman Chmiel/ iStockphoto)*

5,000 operations a second. Yet failures were common, as computer expert Grace Hopper (1906–92) discovered in 1945 when she found a bug—a moth—that flew into the Mark II computer and ruined one of the parts. The term *bug* for a failure or fault did not originate with this incident, since the word had been commonly used years earlier to describe machine defects. But Hopper's discovery did give it a fuller meaning, as she noted in her log: "First actual case of bug being found."

Computer components gradually shrunk in size. An electronics revolution occurred in 1947 when physicists John Bardeen (1908–91), Walter Brattain (1902–87), and William Shockley (1910–89) invented the transistor, a small electrical device that could be used in computer circuits to replace a larger and more energy-consuming component known as a vacuum tube. In 1958, engineer Jack Kilby (1923–2005) developed the *integrated circuit* (IC), an electrical component that contains (integrates) a number of circuit elements in a small "chip." These devices use *semiconductors,* often made with *silicon,* in which electric currents can be precisely controlled. The ability to fit a lot of components on a single circuit decreased the size of computers and increased their processing speed. In 1981, IBM introduced the PC, a small but fast personal computer—this computer was "personal" in that it was meant to be used by a single person, as opposed to mainframe computers that are typically shared by many users.

Mainframes still exist today, although a lot of computing is done with smaller machines. But no matter the size, the basic operation of a computer is similar. A computer stores data—which consists of numbers that need processing or the instructions to carry out that processing—in memory. The central processing unit (CPU) performs the instructions one at a time, sequentially, until the operation is complete. This unit is also known as the processor. Humans interface with the computer by inputting information with a keyboard or some other device, and receive the result by way of an output device such as a monitor or printer.

Most computers today are digital. Instead of operating with numbers that take any value—which is referred to as an analog operation—a digital computer operates on *binary* data, with each *bit* having two possible values, a 1 or a 0. Computers have long used binary data; John V. Atanasoff (1903–95) and Clifford Berry (1918–63) at Iowa State University designed a digital computer in 1940 that used binary data, as did German engineer Konrad Zuse (1910–95) about the same time.

The use of binary data simplifies the design of electronic circuits that hold, move, and process the data. Although binary representation can be cumbersome, it is easy to store and operate on a number as a string of ones and zeroes, represented electrically as the presence or absence of a voltage. Having more than two values would require different voltage levels, which would result in many errors: Brief impulses known as transients, which occur in every circuit, could distort voltage levels enough to cause the computer to mishandle or misread the data. (For the same reason, music in digital format, as on CDs, usually offers better sound quality than analog formats such as ordinary cassette tapes.)

Binary is the computer's "machine language." Because binary is cumbersome to humans, interfaces such as monitors and keyboards use familiar letters and numbers, but this means that some translation must occur. For instance, when a computer operator presses the "K" key, the keyboard sends a representation such as 01001011 to the computer. This is an eight-bit format, representing symbols with eight bits. (Eight bits is also known as a *byte*.) Instructions to program computers

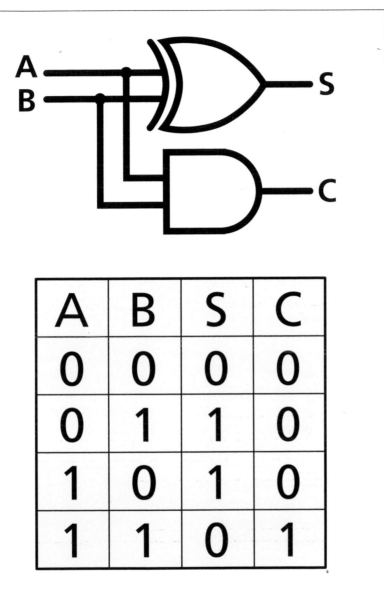

A	B	S	C
0	0	0	0
0	1	1	0
1	0	1	0
1	1	0	1

The gate on top is called an *exclusive or* gate, and the one at the bottom is an *and* gate. The outputs, S and C, are given for each possible input value A and B. This circuit adds the inputs, with S as the sum and C as the carry digit.

must also be in machine language, but programmers often use higher-level languages such as BASIC, C, PASCAL (named after Blaise Pascal), FORTRAN, or others. The instructions of these languages are more human-friendly, and get translated into machine language by programs known as compilers or interpreters.

Circuits in the computer known as logic circuits perform the operations. Logic circuits have components called gates, two of which are symbolized in the figure on page 7. The output of a gate may be either a 0 or a 1, depending on the state of its inputs. For example, the output of an AND gate (the bottom symbol in the figure), is a 1 only if both of its inputs are 1, otherwise the output is 0. Combinations of logic gates produce circuits that can add two bits, as well as circuits that perform much more complicated operations.

Computer engineers build these logic circuits with components such as transistors, which in most computers are etched on a thin silicon IC. Barring component failure, the operations a computer performs are always correct, although the result may not be what the human operator wanted unless the sequence of operations—as specified by the *software* (the instructions or program)—is also correct. If the program contains no errors, a computer will process the data until the solution is found, or the human operator gets tired of waiting and terminates the program. A program that takes a long time to run may not provide the solution in a reasonable period, so the speed and efficiency of programs—and the nature of the problem to be solved—are important. This issue is the subject of chapter 3 of this book. But even an efficient program will tax the user's patience if the computer's circuits—the hardware—are slow. This is one of the reasons why engineers built gigantic machines such as Mark I and ENIAC, as well as the main reason why researchers continue to expand the frontiers of computer technology.

SUPERCOMPUTERS

Special purpose devices such as music players, cell phones, and global positioning system (GPS) receivers contain processors similar to those used in computers, but these processors are designed to perform a single function. Computers are programmable, and the size and complexity of the functions they can perform depend on the speed by which the processor executes the instructions.

One measure of a computer's performance capacity is the clock speed. Operations in most computers are carried out in steps, one instruction at a time, synchronized by a clock. Clock speed or rate is usually given in hertz, which equals one cycle per second, or in multiples of hertz such as a gigahertz, which equals a billion hertz. For example, the author of this book is typing this sentence with a program running on a 3.0 gigahertz computer, which means that the computer can perform 3 billion steps per second. Although this measurement is useful to compare the speed of similar processors, the configuration of a processor is also critical in how fast it can operate. Some processors can do more than others at a given clock rate, which makes comparisons based solely on this number potentially misleading. (And the author cannot type that fast anyway!)

Another measure of a computer's speed is its *floating-point operations per second* (FLOPS). A floating-point is a method of representing real numbers digitally, and operations with these numbers are common in many calculations, particularly scientific ones. Other measures of speed are also used, such as instructions per second (usually identified as millions of instructions per second, or MIPS). None of these measurements can fully characterize the speed of a computer, since factors such as the efficiency of the memory, input and output devices, and other components are critical to the computer's overall performance. But FLOPS and similar speed rates are often used to compare the fastest computers, which are commonly employed in scientific research.

The fastest computers in the world are known as *supercomputers*. A supercomputer is usually larger in size than most computers today, though perhaps not quite as big as the giant mainframes in earlier times. Supercomputers are specially built for speed, so engineers design the memory access and input/output systems to keep the data flowing quickly and at all times. Most other design features, such as light weight and simplicity of operation, are sacrificed. Processing speed is ramped up even more by the use of multiple processors. These processors work simultaneously, which decreases the time required to make a computation but increases the complexity of the computer's design and operation, since many computational tasks are operating in parallel—meaning, at the same time in different circuits.

The term *supercomputer* is relative to other computers presently in existence—because of continual advances in computer engineering, the author's personal computer, which is certainly not a supercomputer, is faster than most machines that were labeled "supercomputers" 20 years ago. The fastest machines at that time, such as computers made by Cray, Inc., could reach speeds of a few billion FLOPS. In November 2008, the world's fastest computer was an IBM machine called Roadrunner, housed at the Lawrence Livermore National Laboratory, in Livermore, California. This supercomputer has achieved a speed of 1,059,000,000,000,000 FLOPS, or 1.059 quadrillion FLOPS (petaflops). Compare that with ENIAC and MARK I of the 1940s, which were capable of a few thousand operations per second, and human beings, of whom the quickest can generate far less than one FLOPS.

In February 1996, a computer beat the reigning chess champion for the first time when IBM's supercomputer Deep Blue beat Garry Kasparov in a game during a chess match, which consisted of a series of six games. Kasparov won the match by winning more games than the computer (four to two). But in a rematch held in May 1997, Deep Blue, with its 256 processors, won the match—the computer defeated Kasparov twice and lost only once, with three games concluding in a draw. Although game playing is not a particularly substantial application of supercomputers, the exercise demonstrates the computational power of these machines. Computers had been winning against unpracticed opponents for some time, but not until Deep Blue could the world's best chess player be defeated. Deep Blue made its moves based on a sophisticated *algorithm*—step-by-step instructions—that examined in great depth the most likely sequences of moves that would be made in the future, given the present positions of the chess pieces.

IBM designed its Blue Gene computers, some of the fastest computers in operation today, to investigate research problems in *protein* structure, *deoxyribonucleic acid* (DNA), genetics, nuclear physics, materials science, and other areas that require extensive calculations. (The genetics research application, which motivated the initial design stages that began in 1999, explains the "Gene" portion of the computer's name. "Blue" comes from Big Blue, a nickname for IBM; although the origin of this nickname is not certain, it may refer to the color of the company's products in earlier times or to the color of its logo.) A computer known as Blue Gene/L has expanded to include 106,496 nodes, each of which is like a computer itself, with memory storage and two processors.

Scientists at Lawrence Livermore National Laboratory have used the Blue Gene/L to simulate the forces and motions of molecules in materials such as crystals and flowing water, providing insight into the behavior and stability of these materials. These calculations help researchers to understand the materials and how to use them in an optimal fashion. But even tiny objects contain thousands of interacting molecules, and a staggering number of calculations must be made in order to determine their behavior. Computer simulations of other complicated processes, such as nuclear reactions or the ways in which amino acids interact to give a protein molecule its shape, may require days or even weeks to finish. Supercomputers are a big help in research, and can defeat chess champions, but even faster speeds are needed to make significant progress on these complex scientific issues.

COMPUTING WITH LIGHT

Blue Gene/L and other fast computers use electricity to represent information and to perform operations on that information. Materials that readily carry an electrical current are called *conductors*. Electrical currents are usually composed of the motion of tiny charged particles called electrons, and although the flow is fast, all conductors resist this movement to a certain extent. The only exceptions to this rule are superconductors, which have no electrical resistance. But all superconductors found so far work only in extremely cold conditions, requiring expensive refrigeration of the device in which they are used. This requirement rules out the use of superconductors in most everyday applications.

When a current flows through an ordinary conductor, heat is generated as the electrons rattle and bump their way through the resistance. High temperatures damage or even ruin small, delicate objects such as electrical components, so every computer must be designed to prevent overheating. Cooling is normally accomplished with convection currents—these currents are not electrical but are streams of air, blown by fans, which carry away excess heat. Large computers such as supercomputers, which draw a lot of electrical current, make extensive use of fans. Increasing the number of processors and equipment elevates the computer's speed but also its temperature as well as its complexity, resulting in a point at which additional processors cause failures.

Suppose that instead of electricity as the medium of operation, a computer uses light. Light is electromagnetic radiation that travels in a vacuum at 186,000 miles per second (300,000 km/s). Not only is light supremely fast, it can travel without generating as much heat as electrons. Light beams can also cross each other, making it possible to have a higher density of paths in a small space.

Electronic computers have transistors and other components made of silicon semiconductors. Electrical currents flowing through these components can be quickly switched on and off. Logic circuits use these switches to control other switches, forming the internal "brain" of the computer. A computer that uses light—an optical computer—would need an optical equivalent of these devices.

An "optical transistor" could be made of a material in which photons—particles of light—control the current. In 2001, University of Toronto researchers K. J. Resch, J. S. Lundeen, and Aephraim Steinberg found a special optical crystal in which they could use a photon to control the flow of other photons (or, in other words, the transmission of a beam of light). But these effects are nonlinear—they do not follow a simple linear interaction in which the outputs are directly proportional to inputs—so they can be difficult to control. The paper, "Nonlinear Optics with Less than One Photon," was published in a 2001 issue of *Physical Review Letters*.

Although an optical computer promises faster speeds—and no moving parts except photons—light is difficult to control and manipulate compared to electrons. But research on the use of optical devices in computers can potentially combine the advantages of light with the advantages of electrons. The resulting computer would be a hybrid, using both light and electricity. This idea is similar to hybrid cars that draw power from both batteries and gasoline engines; batteries alone are generally too weak to power the car, but their contribution reduces the expensive fuel bill and polluting emissions of a gasoline engine.

Researchers such as John E. Bowers, at the University of California Santa Barbara, are trying to replace some parts of conventional computer circuits with lasers. The term *laser* is an acronym—light amplification by stimulated emission of radiation—and the device uses atomic emissions to amplify photons of a single frequency, producing a beam of coherent light. Coherent light does not disperse much, unlike a flashlight beam that spreads out quickly; the energy of coherent light remains concentrated in a small diameter. Lasers are useful for a variety of purposes, including communications, where optical fibers carry signals much more efficiently than copper telephone wires.

Bowers and his colleagues are making lasers out of silicon, and integrating them into computer chips. The use of lasers would boost the rate of data transmission—how fast signals can be moved from one place to another—which is one of the "bottlenecks" of high-speed computing. For instance, the multiple processors of supercomputers must be linked with communication channels so that they can work efficiently on the same problem at the same time. Electrical links are slow.

In 2006, Alexander W. Fang, John Bowers, and their team reported making a laser that could be integrated into a silicon chip. With the addition of elements such as aluminum (Al), indium (In), gallium (Ga), and arsenic (As), the researchers created a laser that gets its energy from, or is "pumped" by, electricity. The laser can produce short, stable pulses of light, which are needed for rapid communication of signals. Research on this laser was reported in "Electrically Pumped Hybrid AlGaInAs-Silicon Evanescent Laser," published in *Optics Express.* As the researchers noted in their paper, "The demonstration of an electrically pumped, room temperature laser that can be integrated onto a silicon platform is a significant step toward the realization of cost-effective, highly integrated silicon photonic devices."

If computer chips can be designed and built using these lasers, the amount and speed of data transmission would be greatly improved, yet the silicon architecture on which most computers are based would remain intact. The lasers would also have many other applications in communications and electronics.

Incorporating optics with electronics would squeeze even more speed out of the semiconductor processors used today. But a limit is fast approaching. Manufacturers such as Intel increase a processor's computational capacity by putting as many transistors as they can on a single chip, packing as many as 1.7 billion transistors in one of their Itanium 2 chips. These components are becoming so small that they are becoming subject to uncontrollable fluctuations, described by principles of physics known as *quantum mechanics.* Although the effects of quantum mechanics may be useful in designing an entirely new breed of computer (discussed below), they ruin engineering applications of conventional electrical circuits by making them unpredictable.

Deoxyribonucleic Acid (DNA)

Until 1944, when Canadian researchers Oswald Avery (1877–1955), Colin MacLeod (1909–72), and American scientist Maclyn McCarty (1911–2005) showed that DNA carries hereditary information, proteins were considered the most likely information carrier. Proteins are strings made from 20 different types of molecules known as amino acids, and are complex molecules capable of assuming many different forms. DNA, in comparison, is a string of only four different types of molecules known as *nucleotides* or bases—A, T, C, and G. DNA seemed too simple to carry the stupendous amount of information necessary to form the blueprint of complex organisms such as humans.

But DNA is well suited for its role. As discovered by British scientist Francis Crick and American biologist James Watson in 1953, the structure of DNA is normally a double-stranded helix, as shown in the figure. Temporary bonds, called hydrogen bonds, between the bases stabilize this configuration. Because of the shape and geometry of the helix and the bases, adenine only bonds with thymine, and cytosine bonds with guanine. The specific base pairing means that if the sequence of one strand of the double helix is AGGTAC, for example, the opposite strand must be TCCATG. Such specificity is critical when DNA is copied, as occurs when a cell divides in two and must provide both daughter cells with DNA. The strands of the helix must be separated in order for enzymes to access the sequence, and each strand forms a template to make the opposing strand—the enzymes insert a T across from an A, a C from a G, and so on, until the original double-stranded helix is copied and two identical double-stranded helices are present.

A single cell in the human body contains 23 pairs of *chromosomes*, which are tightly wrapped packages of DNA.

(continues)

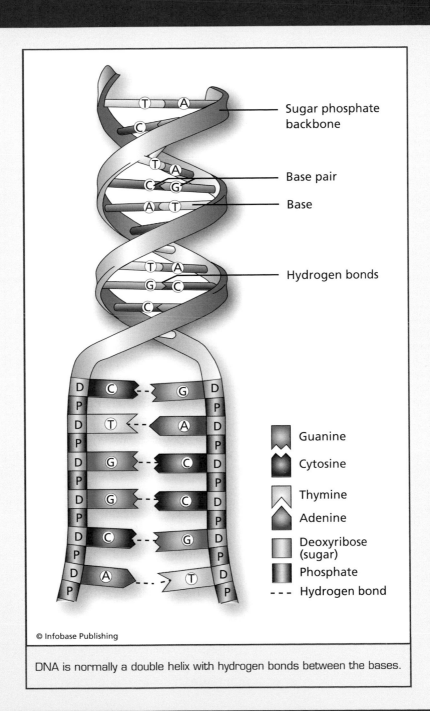

Sugar phosphate backbone

Base pair

Base

Hydrogen bonds

Guanine

Cytosine

Thymine

Adenine

Deoxyribose (sugar)

Phosphate

--- Hydrogen bond

© Infobase Publishing

DNA is normally a double helix with hydrogen bonds between the bases.

(continued)

In total, human DNA contains about 3 billion bases—six feet (1.8 m) of DNA if it were stretched out! Information encoded in the sequence of DNA is passed along from parent to offspring, one chromosome of each pair coming from the mother and one from the father. The base pairing and stability of DNA molecules are important properties for their capacity to carry information, but biologists also exploit these properties when they transfer or clone genes. And the potential exists for other uses of DNA in science and engineering, including computer technology.

In the effort to escape these limits, researchers are exploring alternatives to electronics. Some of these alternatives are slower than electronic computers, but can be used in situations where electrical circuits fail—such as in biological tissues. One option involves learning how to compute with a special biological molecule known as deoxyribonucleic acid (DNA).

COMPUTING WITH MOLECULES

Computers deal with information in binary form. Certain biological systems also deal with information in a certain way, especially the system that transmits information from parent to offspring. In the 1950s and 1960s, researchers such as Francis Crick (1916–2004), James Watson (1928–), and Sydney Brenner (1927–) discovered that this hereditary information is encoded in the sequence of components in long molecules of DNA, a type of nucleic acid. (The term *nucleic acid* derives from the location of many of these molecules—in the nucleus of a biological cell—and their chemical acidity.) As described in the sidebar on page 14, DNA is made of four kinds of covalently bonded units called nitrogenous *bases*—adenine (A), thymine (T), cytosine (C), and guanine (G), arranged in linear order to form sequences known as *genes*. Special enzymes use the sequences of genes to make proteins, which are important molecules performing a variety of critical functions. (For example, enzymes are proteins that speed up the rate of biochemical reactions, such as protein synthesis.)

Erik Winfree, a researcher at California Institute of Technology in Pasadena, and his colleagues have designed logic "circuits" made of DNA. The circuits are composed of mixtures of short DNA sequences instead of electrical components in silicon. Logic gates in these circuits depend on the base-pairing interactions between DNA sequences, which as described in the sidebar are due to temporary bonds between pairs of bases—A and T make one pair, and C and G make another. The molecules are always moving around in solution and will encounter other molecules. When two molecules that have some number of complementary base pairs meet—for instance, one molecule contains the segment ACC at some place in the sequence, and the other molecule contains TGG—the two molecules tend to stick together. But the interaction could be short-lived, because the paired molecules are not tightly bound and are still in motion. If one of the two molecules meets a third molecule that offers a greater number of base matches, the bond between those two will be stronger. As a result, the first pair breaks apart and the two DNA molecules with the larger number of bonds stay together.

Binary values (1s and 0s) in DNA logic circuits are represented by high and low concentrations of molecules. Inputs are composed of certain DNA sequences added to the solution. The outputs are concentrations of certain sequences that result from the base-pair interactions that are due to the inputs. By carefully choosing the sequences, Winfree and his colleagues constructed logic gates such as an AND gate, which outputs a high value (a 1) if both inputs are 1. A high value in this case is a large concentration of a certain sequence that was generated only upon the introduction of a large concentration of two input sequences. These interactions are spontaneous and are rapid enough not to require the presence of enzymes to speed up the process. Georg Seelig, Winfree, and their colleagues published their report, "Enzyme-Free Nucleic Acid Logic Circuits," in *Science* in 2006.

DNA logic circuits are not nearly as fast as electronic ones because electrons and electrical current operate more quickly than base-pairing DNA molecules. But DNA circuits are well-suited for locations in which electronics has trouble, such as the human body. Molecular DNA computation is in its earliest stages, but the development of DNA computers would permit researchers to embed data processing directly in biological or medical applications. Instead of a pill that can only perform one function and fails to adapt to changing circumstances, programmable medical treatments would have the flexibility to respond to a variety of conditions. In addition to improving the range and effectiveness of medicine, computation on the molecular level in biology could also yield tiny but sophisticated robots.

WEARABLE COMPUTERS

While some researchers are working on biological computers, other researchers are engaged in designing and building small computers that will not go inside the body but can be worn on one. Wearable computers will exceed portable or laptop computers in mobility and ease of use.

Sophisticated wristwatch computers found in spy and espionage stories are not yet a reality, but Ann Devereaux, an engineer who works at the Jet Propulsion Laboratory in Pasadena, California, is developing a WARP—wearable augmented reality prototype. A prototype is an early model whose shape and function guide future development. The wearable computer will be voice-activated—engineers have already developed voice recognition systems, and programs are available for most computers—and the user will be able to access communication and information networks, permitting conversations or data downloads. "Augmented reality" occurs when the computer adds information to that available from the user's senses. For example, a wearable computer may receive visual input from a camera, or position information from a global positioning system (GPS), and present the user with a map or coordinates specifying the user's current position, as well as identifying buildings or objects in the user's field of vision.

The Jet Propulsion Laboratory is part of the National Aeronautics and Space Administration (NASA). NASA is interested in wearable computers because astronauts rely on portable computers in places such as the International Space Station or on the space shuttle, but using a laptop requires hands and attention that are usually needed elsewhere. Wearable computers would make the job much easier and safer.

But Devereaux and her colleagues face a lot of design issues. If the wearable computer is too heavy or uncomfortable, it may distract the user at a critical moment. And if the interface between user and computer is awkward, the user may find it difficult or impossible to operate the computer or send and receive data. One version of the prototype consists of a small, wearable box, linked to a headset, microphone, and eyepiece. The eyepiece displays a screen that to the user appears to be projected a short distance away. Testing continues as the researchers try to find the optimal design.

A soldier wearing experimental sensors and computing devices to provide additional information and awareness during combat *(National Institute of Standards and Technology [NIST])*

Wearable computers that are extremely small in size would be useful yet otherwise escape the user's notice. Some researchers are studying methods to make computer components such as transistors out of single molecules. Techniques known as *nanotechnology* operate on scales of about a nanometer—a billionth of a meter, equal to 0.00000004 inches—which is roughly the size of molecules.

Robert A. Wolkow, a researcher at the University of Alberta, in Canada, and his colleagues have succeeded in making a transistor composed of an atom acting on a single molecule. The atom is located on a silicon surface, and has an electrical charge (caused by more or less than the usual number of electrons orbiting the atom's nucleus), creating an electric field. The researchers observed that this field modulated the electrical conductivity of a nearby molecule, or, in other words, changed how well this molecule conducts electricity. By varying the charge of the atom or its position, the researchers could regulate a neighboring molecule's conductivity. This behavior is similar to that of a semiconductor transistor, which switches currents on and off in digital logic circuits.

In addition, the effect is powerful enough to occur even at room temperature, unlike many molecular phenomena that are observable only at extremely cold temperatures (which, in terms of physics, means slow molecular speeds). Paul G. Piva, Wolkow, and their colleagues reported their findings in "Field Regulation of Single-Molecule Conductivity by a Charged Surface Atom," published in *Nature* in 2005.

No one has yet produced a computer processor with such tiny transistors, but the development of molecular-sized logic circuits would increase a computer's speed for a given size. Transistors the size of molecules would achieve much higher transistor densities than can be accomplished with today's ICs—and could fit almost anywhere.

QUANTUM COMPUTER

An increased number of transistors means that the processor can execute more tasks in a given amount of time—recall that supercomputers increase their efficiency by using a large number of processors at the same time. The processors work in parallel, each processor making a small contribution so that the computer can finish the computation more quickly. For the next potential advance in computing technology, researchers and computer engineers are trying to develop a computer that would use parallel processing in an unprecedented manner. This technology would employ the laws of physics that have previously been considered the archenemy of electronic computers—quantum mechanics.

The effects of quantum mechanics present difficulties in electronic computers because unpredictable disruptions tend to occur at small scales and quantities, limiting the miniaturization of circuits. Although molecular transistors offer hope for tiny electrical circuits, linking these components together into usable circuits may prove exceptionally difficult. Yet as disruptive as quantum mechanics can be, it also holds the key to a promising new approach to computers that could be millions of times faster than today's computers. The important quantum mechanical concepts include *superposition* and *entanglement*.

In quantum mechanics, it is impossible to measure with complete accuracy certain states of an object. For example, the position and momentum (the product of mass and velocity) of a particle cannot be known precisely at the same time. This principle is known as the Heisenberg uncertainty principle, named after German physicist Werner Heisenberg (1901–76). The reason for this uncertainty is that tiny particles such as atoms are so small that the act of measuring their position will cause them to jump or move, disrupting their momentum; similarly, measuring momentum will obscure their position. Neither state can be measured simultaneously. This uncertainty is notable in atoms and also holds true in principle for larger objects, but because of their greater mass, the uncertainty is much smaller and generally unnoticeable.

Prior to the act of measurement, the state of an atom or small particle may be any of a number of different values. Common sense suggests that the particle has a state of one value or another, but the laws of quantum mechanics tend to defy common sense. According to quantum mechanics, the particle is in a superposition of these state values—a combination of states—until the act of measurement. Physicists demonstrate this strange behavior by showing that a particle can interfere with itself, causing wavelike patterns on a screen. Interference often occurs when objects with different states combine, but a particle that interferes with itself suggests that it exists in a superposition of states.

Quantum mechanics was first developed in the 1920s. Prominent physicists such as Albert Einstein (1879–1955) criticized the strangeness of these concepts, but more conventional laws of physics had failed to describe the behavior of the atomic world. Austrian physicist Erwin Schrödinger (1887–1961) developed an equation in the 1920s that accurately described the behavior of quantum systems—systems of small particles subject to the concepts of quantum mechanics. These accurate predictions strongly support the theory.

The principle of superposition leads to the idea of a *qubit*—a quantum bit. (Qubit is generally pronounced the same as the word *cubit*.) A bit is normally a single item of information. But in quantum mechanics, an atom can be in a superposition of states—a qubit may be 1 or 0 or some combination of the two. In computer systems in use today, two bits can represent four different values—00, 01, 10, and 11—three bits can represent eight values, and in general, n bits can represent 2^n values. But a single qubit can represent a huge number of values. A small number of qubits could do the work of a large number of bits.

Quantum computers could in principle perform a lot of tasks simultaneously, because the superposition of qubits would allow the computer to work on different problems at the same time, in parallel,

National Institute of Standards and Technology (NIST)

By 1900, only 124 years after declaring independence, the United States had grown and expanded into one of the most prosperous and influential countries in the world. The economy was strong, driven by the power of steel locomotives, steam engines, electricity, and the beginning of automated manufacturing. But the burgeoning economic productivity, along with the complexity of 20th century machinery, created problems when different manufacturers had to work together. Disagreements arose over the units of measurement and how to apply them. For example, one company's "gallon" might not be the same as another company's. Without standards, the parts or containers made by one company did not fit another company's items, resulting in chaos.

On March 3, 1901, the U.S. government chartered the National Bureau of Standards to remedy this problem. Beginning with a staff of 12, the new bureau quickly went to work, improving the standards of length and mass measurements,

similar to the multiple processors of supercomputers. But since the act of measurement disturbs the system—and any readout or printout would constitute a "measurement"—how could the operator obtain the result? The answer is entanglement. Under certain circumstances, the state of two atoms will become entangled; although the states are uncertain—and therefore in superposition—until measured, the states are related mathematically. Measurement of one atom's state lets physicists calculate the state of the other atom. By entangling "read-out" atoms with computational atoms, the user could determine the results of a quantum computer's calculation without disrupting the operation with a direct measurement, and thus avoid ruining the superposition.

and establishing new standards of temperature, time, and light. Time was especially important, as it synchronizes the activities of so many people. How do railroads, radio and television networks, and other time-conscious organizations know what time it is—and keep the same time? The bureau has provided time signals from a radio station, WWV, since 1923. (These days the correct time is also available on the Internet at www.time.gov.) Governing this timekeeping is an atomic clock so precise that it will not gain or lose a second in 60 million years!

The National Bureau of Standards changed its name to the National Institute of Standards and Technology in 1988. Today, the NIST has facilities in Gaitherburg, Maryland, and Boulder, Colorado, and employs about 2,800 scientists, engineers, and staff. The NIST continues to improve measurement technology and standards, helping promote economic and technological progress. Laboratories at the NIST include the Building and Fire Research Laboratory, the Center for Nanoscale Science and Technology, the Information Technology Laboratory, the Manufacturing Engineering Laboratory, the Physics Laboratory, and others.

Although strange and sometimes difficult, the ideas underlying quantum computers are scientifically plausible. But building a computer based on these ideas means working on atomic scales. A research organization that is well equipped for such tasks is the National Institute of Standards and Technology (NIST). As described in the sidebar above, NIST is an agency of the U.S. government devoted to improving scientific methods involved in measurement. NIST studies all kinds of measurement technology, including the difficult techniques associated with tiny particles such as atoms.

NIST researchers have been able to entangle atoms, one of the essential stages of quantum computing. Although particles must be held closely together for entanglement to occur, entangled particles can maintain their relationship over large distances if they happen to move apart. But factors such as stray electric or magnetic fields can disrupt or degrade the process. Dietrich Leibfried and his NIST colleagues have used lasers to hold and entangle two pairs of charged beryllium atoms in an electromagnetic trap—the charged atoms were held in place by electromagnetic forces. With this procedure, the entanglement did not suffer much from distortion, so the researchers were able to "purify" the entanglement and make it less prone to error. By measuring the states of one pair of the entangled atoms, NIST researchers deduced the states of the other pair. While the experiment in itself does not involve computation, it does demonstrate the feasibility of one of the essential ingredients of quantum computers. Rainer Reichle, Leibfried, and their colleagues published the paper, "Experimental Purification of Two-Atom Entanglement," in *Nature* in 2006.

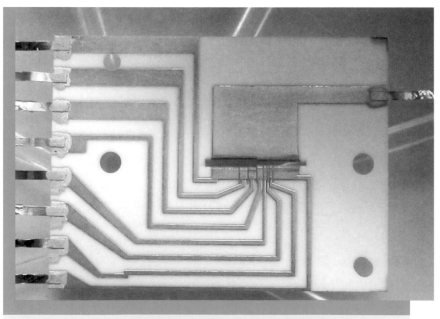

This experimental circuit to test quantum computation includes an ion (electromagnetic) trap (the horizontal opening near the center) made with gold electrodes. *(National Institute of Standards and Technology [NIST])*

Building a computer based on quantum mechanics will not be easy. The NIST's experiment was painstaking and technologically demanding due to the refined techniques, and linking all of these components into a computer will require extraordinary feats of computer science and engineering. But the possibility of boosting computational speeds by millions of times is strong motivation to continue making the effort.

If developed, quantum computers would have revolutionary effects on many aspects of society. For example, information such as credit card numbers and other personal information is kept private during Internet transactions by scrambling the information with methods that are too complex to be broken by today's computers. A quantum computer, however, would be fast enough to crack these schemes in a reasonable amount of time. The loss of security would entail a drastic revision in the way people and companies do business. These procedures are so important that they form an important area of research in computer science, and are covered in chapter 4.

COMPUTERS THAT CAN THINK

The quickness and versatility of advanced computers such as quantum or optical computers make some people wonder if engineers will be able to build a machine that is as smart as a human. *Artificial intelligence* (AI) is a branch of computer science that researches machines or computers that are capable of intelligent activity—playing chess, making decisions, learning, and other activities that require some degree of intelligence.

Some researchers believe it is a matter of time before computers can do almost anything a human can do. For instance, Herbert Simon (1916–2001), the economist and computer scientist who won the 1978 Nobel Prize in economics, maintained until his death in 2001 that computers were or would soon be capable of thinking as well as humans. Other researchers, such as philosopher Hubert Dreyfus, believe that computers are just number-crunchers. Deep Blue, for example, won in chess by running an algorithm that simply generated and searched through a fantastically large number of possible moves in order to find the best one—a "brute force" strategy that differs from the agile mental processes of a human being.

Issues of considerable perplexity cloud the arguments over how successful AI can become, and whether computers actually "think" or are simply elaborate adding machines. To compare computers with humans requires knowledge of both. How do humans think? Psychologists and neuroscientists—scientists who study the brain—have yet to find an answer.

Some AI researchers are trying to install brainlike processes into machines. Two of these methods involve neural networks and computer vision, the subjects of two chapters later in this book. But there remains a question of whether humanlike thought processes can ever be generated by a machine. Further development in AI, as well as continued research in neuroscience, may one day be able to answer this question, although the answer might depend on philosophical viewpoints more than science.

One way of avoiding such debate is by setting up a specific definition or procedure by which the question can be answered. British mathematician Alan Turing (1912–54) devised a test in 1950 to demonstrate computer intelligence. In the *Turing test,* if a machine can fool a person into thinking it is a human, then it must be considered intelligent. Suppose, for example, a person engages in conversation with a number of hidden companions via some mechanism, such as text messaging. An "intelligent" computer will pass the test if its language skills are sufficient to make the person believe it is human.

The Turing test obviously does not examine all the qualities that many people would regard as defining human intelligence. Yet it is a test that has a definite answer. And it is an extremely challenging task—conversation is a basic human skill, yet it is quite sophisticated, and no simple machine is capable of carrying on a conversation without exposing its paucity of skill or knowledge. (In the same way, students who fail to study and do not understand the material are usually unable to fool their professors into getting good grades.) This test is so difficult that many AI researchers today have set less ambitious goals for their projects, although interest in the Turing test remains. Inventor and researcher Hugh Loebner sponsors annual competitions and has offered $100,000 for the first computer developer to build a machine that passes the Turing test. As of May 2009, the money is unclaimed.

CONCLUSION

A spectacular rise in computer speed, portability, and versatility has led to the prominent use of these machines in modern society and technology. Scientific advances and the development of new technology, such as computers based on light, DNA, or quantum mechanics, will allow computers to fill even more roles. Computers whose behavior is almost or entirely indistinguishable from humans may be developed some time in the future.

Another application for the increased power of computers involves a different kind of goal—replicating or imitating a person's environment rather than his or her thinking processes.

Human senses such as sight, hearing, and touch are the sensory "inputs" by which the brain creates human experience. The world has a three-dimensional look and feel. Pictures, graphics, video, and other forms of entertainment or communication are poor mimics of these complex sensations because they do not present enough information. Computer monitors display flat images, for instance; they are only capable of suggesting through various illusions and perspectives the wealth of information added by the perception of depth. But advanced computers may be able to incorporate much more information in their representations—enough information to create a "virtual reality."

Simulations of reality are staples of science fiction, such as in the 1999 film *The Matrix* and its sequels, in which intelligent machines have subdued and exploited the human population by stimulating people's brains and creating an artificial world. More easily achievable simulations involve the presentation of an array of information to the user's senses. Special goggles or headgear present three-dimensional images, and special gloves stimulate tactile (touch) senses of the user's hand. What the user sees and feels mimics the real world, yet it is a computer simulation—a virtual reality.

Replicating rich sensory experience requires a lot of computational power. Even more power is needed if the users are allowed to interact with the virtual environment as they would with the real one. Movement of the head, for instance, changes visual perspective. If a virtual reality user moves his or her head, the images presented by the computer should change accordingly. Quick responses to these changes require extremely fast computers, otherwise there is a lag between the time the user moves and the updated image. Such lags ruin the "reality" of the simulation. (And the lags can make the user quite dizzy.)

Other simulations attempt to create a "world" for a number of observers at the same time. At Iowa State University in Ames, researchers have built a 1,000-cubic-foot (28-m³) room—10 feet (3.05 m) to a side—in which detailed images and sound create a rich sensory experience for its occupants. The room equipment was upgraded in 2007, and consists of 96 computer processors that control 24 digital projectors and an eight-channel surround sound system. The simulator, known as C6, has six screens—one for each of the four walls, one for the floor, and one for the ceiling—to envelope the viewers. Being inside C6 is much closer to reality than watching a small screen.

The potential uses of these simulations go beyond a breathtaking experience such as flying a fast jet or walking on the Moon. Recreating situations or objects in rich detail allows inspectors to examine structures for damage, young pilots to train in realistic flight situations, and many other applications. Technology, such as flight simulation, is improving with the use of virtual reality.

But the need for computer speed continues to be a limiting factor. As computers get faster, their role in society will keep expanding. There would seem to be few if any limits to how far this frontier of computer science can go.

CHRONOLOGY

fifth century B.C.E.	Although various mechanical devices to aid counting had probably already been invented, counting boards and early versions of the abacus begin to be used at least by this time, and perhaps earlier, in China and the Near East.
1623 C.E.	German scientist Wilhelm Schickard (1592–1635) designs a prototype mechanical calculator.
1642	French philosopher and scientist Blaise Pascal (1623–62) builds a functional adding machine.
1822	British engineer Charles Babbage (1791–1871) begins to design and build the "difference engine," a powerful mechanical calculator that was never completed.

1832	Babbage patents an "analytical engine," a general purpose machine that would have been the world's first computer, had it been built.
1890	American engineer Herman Hollerith (1860–1929) develops a tabulating machine that helps the United States government finish the 1890 census much faster than it had the hand-tabulated 1880 census.
1896	Hollerith founds the Tabulating Machine Company, which later became International Business Machines (IBM).
1937	American mathematician and engineer Claude Shannon (1916–2001) develops fundamental concepts of digital circuits.
1940	John Atanasoff (1903–95) and Clifford Berry (1918–63) at Iowa State College (now University) build an electronic digital computer prototype. German engineer Konrad Zuse (1910–95) builds a programmable digital computer.
1944	American engineer Howard Aiken (1900–73) and colleagues finish the Harvard Mark I (also known as the Automatic Sequence Controlled Calculator), one of the earliest programmable electronic computers.
	British engineers perfect electronic, programmable machines known as Colossus computers.
1945	American engineers J. Presper Eckert (1919–95) and John Mauchly (1907–80) finish the Electronic Numerical Integrator and Computer (ENIAC), an electronic computer capable of running stored programs, at the University of Pennsylvania, in Philadelphia.
1947	American physicists John Bardeen (1908–91), Walter Brattain (1902–87), and William Shockley (1910–89) invent the transistor.

1958	Jack Kilby (1923–2005), an engineer working for Texas Instruments, develops the integrated circuit (IC).
1981	IBM launches the PC (personal computer), a small computer for a single user.
1982	American physicist Richard Feynman (1918–88) suggests the possibility of a computer based on principles of quantum mechanics—a quantum computer.
1997	IBM's supercomputer Deep Blue defeats reigning champion Garry Kasparov in a chess match.
2000s	Researchers at the National Institute of Standards and Technology along with their colleagues experiment with techniques to make quantum computers.
2005	Researchers Paul G. Piva, Robert A. Wolkow, and their colleagues succeed in making a transistor composed of an atom acting on a single molecule.
2006	Researchers Georg Seelig, Erik Winfree, and their colleagues construct logic gates using DNA.

FURTHER RESOURCES
Print and Internet

Babbage, Charles. *Passages from the Life of a Philosopher.* London: Longman, Green, Longman, Roberts, & Green, 1864. Babbage, a computer pioneer, describes his life and discusses his thoughts and ideas on computing technology.

Fang, Alexander W., Hyundai Park, Oded Cohen, Richard Jones, Mario J. Paniccia, and John E. Bowers. "Electrically Pumped Hybrid AlGaInAs-Silicon Evanescent Laser." *Optics Express* 14 (2006): 9,203. Available online. URL: www.ece.ucsb.edu/uoeg/publications/

papers/fang_06_opex.pdf. Accessed June 5, 2009. The researchers report making a laser that can be integrated into a silicon chip.

Hally, Mike. *Electronic Brains: Stories from the Dawn of the Computer Age.* London: Granta Publications, 2005. This book describes the advent of modern computers in the 1940s and 1950s, including ENIAC, UNIVAC, and IBM, and tells the stories of the people who made them.

Hsu, Feng-Hsiung. *Behind Deep Blue: Building the Computer That Defeated the World Chess Champion.* Princeton, N.J.: Princeton University Press, 2002. The author, a member of the team that created IBM's Deep Blue, chronicles the development of the first computer to defeat a reigning world chess champion.

Johnson, George. *A Shortcut through Time: The Path to the Quantum Computer.* New York: Knopf, 2003. With simplified and easily understandable language, this book gives an overview of computer basics, and describes the revolution that would be created by the development of quantum machines.

Piva, Paul G., Gino A. DiLabio, Jason L. Pitters, Janik Zikovsky, Moh'd Rezeq, Stanislav Dogel, et al. "Field Regulation of Single-Molecule Conductivity by a Charged Surface Atom." *Nature* 435 (2 June 2005): 658–661. The researchers succeed in making a transistor composed of an atom acting on a single molecule.

Reichle, R., D. Leibfried, E. Knill, J. Britton, R. B. Blakestad, J. D. Jost, et al. "Experimental Purification of Two-Atom Entanglement." *Nature* 443 (19 October 2006): 838–841. The researchers use lasers to hold and entangle two pairs of charged beryllium atoms in an electromagnetic trap.

Resch, K. J., J. S. Lundeen, and A. M. Steinberg. "Nonlinear Optics with Less than One Photon." *Physical Review Letters* 87 (2001): 123,603. Available online. URL: arxiv.org/abs/quant-ph/0101020. Accessed June 5, 2009. The researchers describe a special optical crystal in which they can use a photon to control the flow of other photons.

Scientific American: Understanding Artificial Intelligence. New York: Warner Books, 2002. Containing a set of accessible essays written by experts and originally published in *Scientific American,* this book explores advanced topics such as robots, computational logic, computing with molecules, and artificial intelligence.

Seelig, Georg, David Soloveichik, David Yu Zhang, and Erik Winfree. "Enzyme-Free Nucleic Acid Logic Circuits." *Science* 314 (8 December 2006): 1,585–1,588. The researchers construct elementary logic circuits using DNA.

Stokes, Jon. *Inside the Machine: An Illustrated Introduction to Microprocessors and Computer Architecture.* San Francisco: No Starch Press, 2007. Starting with basic computing concepts, this book proceeds to describe the details of how processors such as the Intel Pentium work.

Web Sites

Computer History Museum. Available online. URL: www.computer history.org. Accessed June 5, 2009. Established in 1996 and located in Mountain View, California, the Computer History Museum houses thousands of artifacts relating to the rise and development of the computer. The museum's Web site offers online exhibits chronicling semiconductors, the Internet, computer chess, and microprocessors.

IBM Research. Available online. URL: www.research.ibm.com. Accessed June 5, 2009. Scientists employed by this large and well-known computer company investigate all aspects of computers and computer technology. This Web site offers information on IBM's main research areas such as computer science, materials science, physics, and mathematics, and has links to IBM laboratories at Almaden, California; Austin, Texas; Yorktown Heights, New York; Zurich, Switzerland; Beijing, China; Haifa, Israel; Tokyo, Japan; and New Delhi, India.

Massachusetts Institute of Technology. The Media Lab. Available online. URL: www.media.mit.edu. Accessed June 5, 2009. Located in Cambridge, the Massachusetts Institute of Technology (MIT) is academically strong in many branches of science and technology, particularly computer science. The Media Lab conducts research and trains students in an interdisciplinary environment that incorporates such fields as nanotechnology, computer interfaces, robotics, wearable computers, and much more. The laboratory's Web site describes ongoing research projects and recent results.

National Institute of Standards and Technology. Quantum Information. Available online. URL: qubit.nist.gov. Accessed June 5, 2009. The Physics Laboratory at NIST conducts a lot of pioneering research in the field of quantum computation. Included on this Web site are an overview of quantum information, descriptions of the laboratory's current projects, and some of the laboratory's publications.

Top 500 Supercomputing Sites. Available online. URL: www.top500. org. Accessed June 5, 2009. This Web site maintains a list of the 500 most powerful computers in the world, based on information gathered by surveys and in publications. The site also keeps historical top 500 lists dating back to June 1993 to track the evolution of supercomputers.

Artificial Neural Networks— Computing with "Brainy" Technology

Computers outperform humans in many tasks. Although humans must write the instructions, once the program is up and running, a computer can perform arithmetic or sort a list in a fraction of the time a person would require to do the same job. As described in chapter 1, the most advanced computers today are trillions of times faster than humans in certain tasks, and IBM's supercomputer Deep Blue defeated Garry Kasparov, the reigning chess champion, in a 1997 chess match.

But even the fastest computers cannot outperform humans in all tasks. Computers excel at tasks requiring a large number of simple operations, but unlike humans, computers are not yet generally capable of making new discoveries. The human brain is an astonishingly complex organ composed of billions of cells; one type of cell, called a *neuron,* communicates with other neurons to create vast networks. The complexity, adaptability, and information-processing capacity of these neural networks provide humans with the intelligence to conduct experiments, test scientific theories, formulate general principles, learn new things, and write computer programs. A computer can only carry out its instructions. Computers are able to run complicated programs, but the program must consist of a sequence of simple instructions, and a computer's processor can only follow these instructions—it does what it is told to do, and nothing more. Deep Blue won its chess match by performing billions of simple calculations that evaluated the outcome of potential moves.

Artificial intelligence (AI) is a branch of computer science aimed at creating machines capable of showing a certain degree of intelligence. The ultimate goal of AI is a computer that can think like a person. One option to reach this goal would be to give computers a "brain" that is similar to a human brain. Many AI researchers who pursue this option have started tinkering with artificial neural networks, which are not biological though they are based on the operating principles of the brain. This chapter describes artificial neural networks—often just called neural networks—that can learn on their own and solve new or complex problems, even when human programmers are unable to specify the exact instructions necessary to do so.

INTRODUCTION

Artificial neural networks were not possible until scientists had some idea about the biological neural networks in the brain. Neurons are enclosed in a membrane and are tiny, having a cell body with a diameter of about 0.02–0.06 inches (0.05–0.150 cm) and a long, thin projection called an *axon*.

Detailed study of neurons began in 1873, when Italian researcher Camillo Golgi (1843–1926) developed a method of staining the cells so that they could be easily viewed in microscopes. Neurons, like most cells, are mostly transparent, and they are tightly packed together, making these small objects nearly impossible for scientists to see and study even under microscopic magnification. Golgi's method involved a dye consisting of silver nitrate, which some (though not all) neurons take up. The dye stained these neurons and made them stand out against a background of unstained cells. (If the dye had stained all the cells, the result would have been a uniform field of color—as useless as the original, transparent condition, because researchers could not have studied individual cells.) Why some but not all neurons take up this dye is still not well understood, but the method gives scientists a good look at these important cells.

Using Golgi's technique, Spanish anatomist Santiago Ramón y Cajal (1852–1934) suggested that neurons process information by receiving inputs from other cells and sending outputs down the axon. Cajal's theories proved to be mostly correct. Neurons send and receive

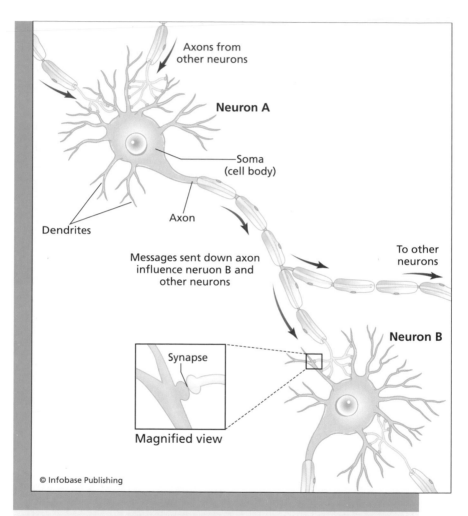

Neuron A, shown above, receives synaptic inputs from many neurons. The soma is the cell body, and dendrites are extensions that often receive synapses. One synapse is shown magnified; in this synapse, neuron B is the postsynaptic neuron. Neuron A sends its axon to make synapses with neuron B, as shown on the right side of the figure. In these synapses, neuron A is the presynaptic neuron and neuron B is postsynaptic.

information from cell to cell by way of small junctions called *synapses,* named by British physiologist Sir Charles Sherrington (1857–1952) in 1897. As shown in the figure, synapses are usually formed between the axon of the sending neuron—the *presynaptic neuron*—and a dendrite or cell body of the receiving neuron—the *postsynaptic neuron.* The figure illustrates the anatomy of a neuron and its synapses.

Information in the brain has a different form than it has in a computer or in human languages such as English. Neurons maintain a small electrical potential of about -70 millivolts (a millivolt is a thousandth of a volt)—the interior of a neuron is about 70 millivolts more negative than the outside. This small voltage is only about 1/20th the voltage of an ordinary flashlight battery and is not powerful by itself (though some animals such as electric eels can combine the small potentials produced by their cells to generate a powerful shock). More important is a neuron's ability to change its voltage briefly, causing a voltage spike that lasts a few milliseconds. The spike is known as an action potential.

Neurons transmit information in the form of sequences of action potentials. An action potential travels down an axon until it arrives at a special site called an axon terminal, which is usually located at a synapse. In most synapses, the spike causes the presynaptic neuron to release molecules known as neurotransmitters that cross the synaptic gap and attach to a receptor in the postsynaptic membrane. This activates the receptor, which sets certain biochemical reactions into motion and can slightly change the potential of the postsynaptic neuron. Neurons are continually receiving these synaptic inputs, usually from a thousand or more neurons, some of which slightly elevate the neuron's potential and some of which depress it. A neuron will generally initiate an action potential if its voltage exceeds a threshold, perhaps 10 or 15 millivolts higher (more positive) than its resting potential of -70 millivolts. In this way, neurons are constantly "processing" their inputs, some of which are excitatory, tending to cause the neuron to spike by pushing it closer to the threshold, and some of which are inhibitory, making it more difficult for a neuron to spike by dropping the potential farther away from the threshold. The result of this processing is the brain activity responsible for all the intelligent—and sometimes not so intelligent—things that people do.

Vision, for example, begins when special cells in the eye called photoreceptors absorb light. Other cells convert the light signals into trains of action potentials that represent the dark and bright areas making up the

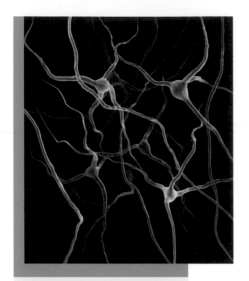

Drawing of a network of neurons showing individual cells—networks in the human brain are generally much more densely populated *(Gary Carlson/Photo Researchers, Inc.)*

image. Dozens of neural networks, distributed over vast areas in the brain, process this visual information, extracting information such as the number and type of objects and the color and motion of these objects. At some point—scientists are not sure how and where—the person perceives and becomes consciously aware of this visual information.

Information processing in the brain is much different than in an ordinary computer. A computer generally operates on binary values using digital logic circuits to transform data. Each processor in a computer works serially, one step at a time. In the brain, information processing occurs in parallel. Millions of neurons are working at the same time, summing their synaptic inputs and generating more or fewer action potentials. This activity is sometimes called parallel distributed processing, which refers to the simultaneous parallel operations distributed over a broad area.

The parallel nature of information processing in the brain is the reason it can work so quickly. Computers are much faster in arithmetic, but the brain's primary function is not to add or subtract numbers quickly. The brain evolved to analyze sensory inputs—vision, hearing, smell, taste, and touch—and extract vital information concerning food and predators. Neural networks in the brain can interpret an image more rapidly and accurately than any computer program, for example. Each neuron behaves like a little processor, contributing its portion to the overall computation. Supercomputers gain speed by using a lot of processors working in parallel, but the brain has approximately a trillion neurons, which gives it a computational capacity greatly exceeding computers for jobs it evolved to perform.

REAL NEURONS AND MODEL NEURONS

In the 1940s, when researchers such as Konrad Zuse, John Atanasoff, and others were designing and building the first digital computers, as described in chapter 1, Warren McCulloch (1898–1968) and Walter Pitts (1923–69) were studying the computational ability of neurons. McCulloch, at the University of Illinois College of Medicine, and Pitts, at the University of Chicago, had several goals—to discover how brains can think, and to build a device capable of imitating this process.

McCulloch and Pitts designed a simple model of a neuron. A model is generally a simplified version of a process or an object under study. The model may be built on a smaller scale, such as the model airplanes that engineers use to test aeronautical performance in a wind tunnel, or the model may only exist on paper, in which case researchers study its theoretical properties or behavior. Most importantly, a model must incorporate all the essential features of the object or process under study, and discard the rest—this is how the model simplifies the research effort while maintaining its accuracy. For example, engineers who are studying the aerodynamics of an airplane's wing must use a model that has the precise shape of the wing (although probably on a reduced scale), but other features of the airplane, such as the landing struts, are irrelevant.

A real neuron is a complex biological cell. Many of a neuron's properties, including a large number of biochemical reactions, are vital but not directly related to its computational function. McCulloch and Pitts abstracted what they believed were the critical features involved in computation, and included only these features in their model neuron. The model neuron sums a number of inputs and produces a binary output—the neuron either "fires" or it does not. (This output resembles the binary values represented in digital computers—a 1, which means the neuron fires, and a 0, which means the neuron is silent.) A decision to fire or not is based on a threshold. If the summed inputs exceed the threshold, the neuron fires; otherwise it does not.

Using their model neurons, McCulloch and Pitts showed how logic circuits could be implemented with neural networks. Logic circuits form the basis of a computer's operations as it processes information in binary form. For example, computers use these circuits, made with electronic elements such as transistors, to add numbers as well as other, more complex tasks. The simple, artificial neurons of McCulloch and Pitts could also be used to make such circuits. F. H. George, a professor at the University of

Bristol in England, enthusiastically wrote in *Science* in May 30, 1958, that McCulloch and Pitts broke new ground "when they saw that the brain is rather like a machine that can be described by a mathematical system that has values of its variables of 1 or 0. . . ."

But is this how the brain actually works? Scientists realized that networks of such simple neurons do not necessarily compute in the same way as the brain does. Binary neurons, while able to duplicate the processing of binary computers, are not always representative of real neurons in the brain. Neuroscientists have discovered that neurons encode information in a variety of ways, not just a binary-valued spike or not. For instance, neurons encoding visual information may fire spikes at a faster rate as the image increases in brightness.

But the model of McCulloch and Pitts demonstrated the computational ability of simple neuronlike elements. This model led the way to more elaborate networks, capable of even greater computation, as well as computations that more closely resemble those of the brain.

GETTING CONNECTED—NEURAL NETWORKS AND SYNAPSES

Cornell University researcher Frank Rosenblatt (1928–71) began working on neural networks that he called perceptrons in the late 1950s. As illustrated in the figure, a simple type of perceptron is a network of model neurons consisting of one group, or layer, of neurons receiving the input, which is similar to the process of feeding data into a computer. The neurons of the input layer have a connection with each neuron of the output layer, by which the input neuron influences the state of the recipient (the postsynaptic neuron). These connections are called synapses because they perform the same function as synapses in the brain. The connections have a *synaptic weight,* meaning they do not necessarily influence the recipient neuron equally—some synapses may have a greater weight, or impact, on the postsynaptic neuron. Output neurons project the result of this computation to the user. Each neuron sums its inputs and decides whether or not to fire a spike based on the sum. The decision to fire can be triggered by exceeding a threshold, but decisions can also be made with more complicated methods.

Rosenblatt showed that perceptrons have the capacity for associative memory, meaning that a specific output is associated with a specific input. In other words, the perceptron will produce that output when given the correct input. This procedure is similar to the memory triggered in a person's brain by the image, say, of a friend; a person sees the face of a friend, and the name of the friend comes to mind. In the perceptron, the input would be a *digital image* of a face, and the output would be a representation of the name. (Digital images are composed of little points called *pixels,* the intensity and color of which is represented by numbers.) For example, with a digital image of James Bond as input, a perceptron may output 000000111, which is a binary number representing the decimal number 007. Rosenblatt studied the theoretical operation of perceptrons, and in 1960 helped build a machine, the Mark I Perceptron, which was a perceptron made with electrical and mechanical parts.

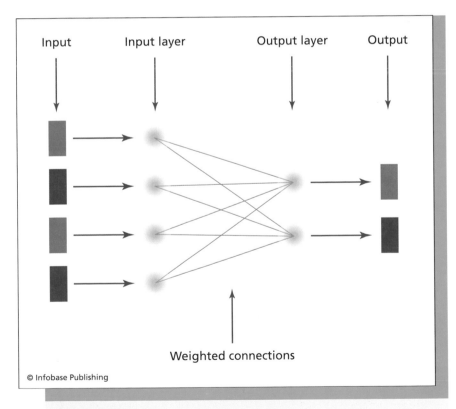

© Infobase Publishing

A simple perceptron, with an input and output layer that are connected with synapses having a certain synaptic weight

The parallel nature of neural network computation is evident because the inputs are processed all at once, rather than serially (one at a time). Parallel computation is much faster.

Rosenblatt's work generated a lot of excitement, as McCulloch and Pitts had done earlier. In August 22, 1969, Allen Newell, a researcher at Carnegie Mellon University in Pittsburgh, Pennsylvania, noted in *Science* that Rosenblatt's perceptrons "became both popular and controversial, since these devices were viewed by some as having remarkable powers of self-organization and as being the first true toehold into the development of really intelligent devices."

Another advantage of neural networks such as perceptrons is redundancy. In the brain, the loss of a few neurons will not eliminate a memory or a specific type of calculation. While each neuron in the network makes a contribution, there is some overlap, or redundancy, so that the absence of a neuron or two does not affect the result. Redundancy is important because neurons continually die in the brain, but memories are relatively unaffected. (It is only in diseases such as Alzheimer's disease, which kills a substantial fraction of neurons, that a loss of memory and mental capacity becomes evident.)

Neural networks in the brain are also able to produce accurate results even if the input is "noisy" or only partially revealed. For example, people can recognize a friend even after the friend has gotten a haircut or is wearing a baseball cap. Considering the variety of angles from which a person can be viewed, as well as the varying distances, the brain must be able to associate a name and identity to a face that can have a strikingly different number of actual appearances. Perceptrons can also do this. An input having, say, 20 binary values, may be incorrect in one or two values—a 1 that should be a 0, and vice versa—yet the perceptron may still generate the correct output. In such a case, neurons processing the correct portion of the input are able to override neurons that are "fooled" by the wrong inputs. As a result, the network settles on the correct output.

Despite the use of biological principles and neural terminology, simple perceptrons are not identical to the complex networks of the human brain, nor do they function the same way. Still, perceptrons suggest a method by which a computer may be able to imitate some of the advantages of human brainpower.

But there are limitations, as pointed out in a 1969 book, *Perceptrons*, written by researchers Marvin Minsky and Seymour Papert. The authors showed how a perceptron with only an input and output layer cannot compute certain functions, such as a logic function known as an "exclusive or." (An exclusive or gate is also known as a "not equal to" function, since it returns a 1 only when one input is 0 and the other input is 1—in other words, when its inputs are not equal.) This made perceptrons appear much less useful, and led to a lull in the development of neural networks as computational devices—until researchers discovered methods to create more advanced networks.

TRAINING A NEURAL NETWORK

Boston University researcher Stephen Grossberg was one of the few researchers who continued working on neural networks in the 1970s, using advanced mathematics to study network properties and laying the foundations for further studies. Grossberg and other researchers explored neural networks that could eliminate the limitations and criticisms asserted by Minsky and Papert. McCulloch and Pitts showed earlier that their networks could generate any logic function, including the exclusive or. Extra layers added to perceptrons—in addition to the input and output layers—enabled perceptrons to perform these functions. The problem was that these extra layers were hidden between the input and output, and Minsky and Papert claimed that it would difficult to adjust the properties of "hidden" layers. They believed that hidden layers would make the network too complicated to control.

But in 1974, Paul Werbos published a process by which multilayered perceptrons could be trained. The publication was Werbos's Ph.D. thesis, which documented the research Werbos conducted as a graduate student at Harvard University in order to obtain his doctorate degree. The method of Werbos, called *backpropagation,* adjusts the synaptic weights of the network.

Other researchers had already been adjusting the properties of neural networks by changing the weights of synapses. By raising or lowering certain weights, the influence of certain neurons and synapses can be increased or decreased, since the influence these neurons have on other neurons will change accordingly. But backpropagation is a general method to train a neural network, including one with hidden layers.

Backpropagation occurs in a series of steps. The method is not automatic, since it requires a "teacher" or "supervisor"—a person who knows what the network's output should be for a given input. In other

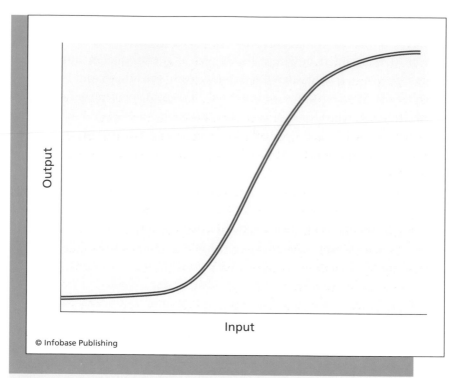

This curve, often called sigmoidal or S-shaped, shows the neuron's output for a given input.

words, the supervisor knows the state—firing or not firing—that the output neurons should have, given the state of the input neurons.

In the first step, the input layer of the network is given a specific input, which the neurons process in the same way as described earlier—firing or not firing, depending on the sum of their weighted synaptic inputs. The output of the network, as given by the output layer, will usually be different initially than what is desired because the network has not yet been trained. Some of the output neurons may be at a 1 (firing) when they should be 0, and vice versa. Neurons that made synapses with these output neurons are what caused the error, so the supervisor calculates the error and lowers the synaptic weight of neurons that led the output astray, so to speak. The correction of the errors therefore propagates backward in terms of synaptic inputs, from output to the hidden layer or layers.

Although the mathematics of this procedure are too complicated to present here, the basic idea is that the adjustments result in fewer errors the next time the input is given. These steps are repeated until the errors are reduced to an acceptable level or eliminated entirely.

Many neural networks use backpropagation, along with somewhat more complicated methods of determining whether a neuron will fire a spike or not. The simple threshold of McCulloch and Pitts is often replaced with a function whose curve looks like a steep hill, as shown in the figure. Rather than having an output of a 0 or 1, such neurons will have a range of values; the output will be low when the inputs are low, but rise steeply when the inputs increase in value.

Adjusting the behavior of artificial neural networks by changing synaptic weights is not just a wild idea dreamed up by computer scientists. Experimental evidence shows that synapses in the brain can vary in strength—sometimes a presynaptic neuron has more or less influence on the postsynaptic neuron, as measured by the effect on the postsynaptic neuron's electrical potential.

There is also some evidence that these variations are associated with learning and memory. When a person learns a new skill or a new bit of information, something in the brain must change—otherwise the brain would not be able to retain the new knowledge. In experiments with animals such as certain marine invertebrates, learning something new results in observable changes in the synapses of neurons that process the information. Although invasive experiments on humans cannot be performed, basic similarities in the nervous systems of all animals mean that the physiological changes associated with learning are probably similar. Adjustments of synaptic weights in artificial neural networks do not necessarily reflect identical biological processes, but the idea has a biological basis.

Excitement over neural networks burgeoned in 1982 when California Institute of Technology physicist John Hopfield published a paper, "Neural Networks and Physical Systems with Emergent Collective Computational Properties," in the *Proceedings of the National Academy of Sciences*. Hopfield showed that a certain type of neural network, later called a Hopfield network, could settle into the correct states automatically. The computation is collective in the sense that all neurons participate, and the complex properties of the network "emerge" or arise from the collective activity of a network of simple neurons.

Instead of having different layers, a simple Hopfield network consists of a group of fully connected neurons (each neuron makes a synapse with all of the others). The "input" of the network is the state of all its neurons at the initial stage of the computation. Each neuron receives a weighted synaptic projection from the rest of the neurons, and after the summation and decision to fire or not, the neuron's state may change. This is the end of one step or cycle. Since all the neurons are connected, a change in state of any of the neurons will change the synaptic inputs to all neurons. The value of the weighted synaptic inputs must be recalculated, and the state of each neuron updated; this is the next cycle. Cycles occur repeatedly until no more changes occur and the network settles down into a specific, stable state. This state is the "output" that is associated with the given input. Hopfield showed mathematically that under certain conditions the network would always find a stable state, rather than continually changing.

Hopfield networks have associative memory—the neural network associates a specific output (the stable state) with a specific input (the initial state, set by the user). Because of the collective nature of the computa-

Hebbian Learning

Canadian psychologist Donald O. Hebb (1904–85) obtained his Ph.D. from Harvard University in 1936, at age 32, and was interested in how people learn and remember. Earlier researchers such as Cajal had suggested that synapses may underlie the brain mechanisms of learning and memory, and Hebb wondered what kind of synaptic activity might be involved. In 1949, Hebb published a book, *The Organization of Behavior*, in which he sketched some of his theories.

Hebb was struck by how the mind associates different objects to form a pattern or a memory. For example, the sight of a trophy might remind a person of the event or game in which the trophy was won, and the memory of this event, in turn, may spark other memories, such as the congratulations of parents and friends, and so on. Memories often form a chain of associations, with one memory invoking the next. Perception functions in a similar way, with a stimulus, say an apple, invoking a chain reaction, perhaps leading to the realization that one is hungry. But no one knows exactly how such associations form in the brain.

tion, the network can "retrieve" this memory even if the input is slightly different. Should a few neurons start in the wrong state, the network will still settle into the appropriate state, although it may take longer than usual. As with the human brain, the memory of Hopfield networks can be "jogged" with only some number of "clues" or "reminders."

The number of memories a network can store depends on the number of neurons. A greater number of neurons in a Hopfield network allows more associations to be correctly performed, but eventually a limit is reached. Hopfield showed that in a network of N neurons, "About 0.15N states can be simultaneously remembered before error in recall is severe," as he wrote in his 1982 article. For example, a Hopfield network containing 100 neurons can store about 15 associations; fewer if more stringent accuracy is needed.

One of Hebb's most important hypotheses is that the connections between neurons increase if their activity is correlated. Suppose that neuron A makes a synapse on neuron B, though the connection is weak. (In other words, the weight of this synapse is nearly zero.) Note that A's synaptic input to B is just one of many inputs to B. If neurons A and B are simultaneously active over some period of time, Hebb hypothesized that the strength of the connection between these neurons will grow—"some growth process or metabolic change takes place in one or both cells such that A's efficiency, as one of the cells firing B, is increased." This increase in strength means that in the future, the firing of neuron A will be more important in B's decision to fire or not, so A and B will be more strongly correlated. The association has been learned.

The strengthening of connections caused by correlated activity throughout the network will result in a chain of associations by which one item, be it a perception or memory, invokes others. This hypothesis, called Hebbian learning, has influenced computer scientists exploring artificial neural networks, as well as scientists who study the brain. Although no one knows how much of a role Hebbian learning plays in brain mechanisms, certain synapses do sometimes follow this principle.

Artist's conception of a neural network superimposed on a schematic of a computer circuit—neural networks could eventually replace many of today's computers *(Alfred Pasieka/Photo Researchers, Inc.)*

Suppose that one wants to assign a specific output in a Hopfield network to a specific input. How is a Hopfield network trained? Unlike the backpropagation method, training a Hopfield network does not require a propagation of error corrections but instead depends on the correlation of the neural activities. Hopfield networks compute iteratively (repetitively, in a series of steps or cycles), and the output appears when each neuron settles into a stable, unchanging state. For these outputs, some of the neurons will be positively correlated—for instance, their value will be high or 1—and some neurons will be negatively correlated, so that their values differ (one may be high and another low).

To "memorize" certain associations, correlations between neurons are calculated and the synaptic weights set accordingly. If two neurons are positively correlated most of the time, their synaptic weight has a large positive value, so that they will positively influence one another. (As mentioned earlier, all neurons are connected with one another in Hopfield networks. The synaptic weights are generally symmetric—the synaptic weight for neuron A's projection to B is the same as neuron B's projection to A.) Synapses between negatively correlated neurons have a negative value, so these neurons tend to have the appropriately opposite behavior. For instance, a neuron with a lot of activation will suppress the other, giving it a low value. This principle of learning or training, based on correlation, is called Hebbian learning, and is discussed further in the sidebar on page 46.

Neural networks such as perceptrons and Hopfield networks store memories in a vastly different way than most computers today normally do. A computer's memory stores data at specific locations. When a particular piece of data is needed, the computer's circuitry looks up the address and retrieves the information. Neural networks store memories in the collective computation of the neurons, as determined by synaptic weights, and retrieve the memories when provided a specific input or clues. This is extremely important when the input is not exact.

Consider, for example, reading someone's handwriting. While the basic shape of letters is the same, everyone has a slightly different style. Conventional computers have a lot of trouble reading handwriting, but neural networks do not.

INTELLIGENT CHARACTER RECOGNITION—NEURAL NETWORKS THAT READ HANDWRITING

Optical character recognition (OCR) is a process by which a computer can convert the text of a typewritten or printed document into a file stored on the computer. A piece of equipment called a scanner creates an image of the document, and OCR software running on the computer identifies the characters of the text. The result, if all works well, is a digital file in the computer containing the same text. This allows a person to send, edit, or store the text electronically, without having to use the keyboard to enter the text.

All does not always go well, however. OCR works only for documents in which the print is dark and clean, and the font—the shape of the characters—is recognizable. Otherwise the computer program makes mistakes. The author of this book recently used OCR in an attempt to produce an electronic copy of an old, faded document. On the first page, the computer converted the phrase "It should be herewith stated" to the phrase "It shoot be hero ilk stated."

The problems are compounded when the text is not printed but handwritten. Some people have better penmanship than others, but the characters of even the most careful writers display a certain amount of variation. Unless the characters are exactly the same, the performance of OCR programs drops considerably.

Neural networks offer a solution to this problem. Because neural networks do not need the full pattern to recognize an input, these devices can arrive at the correct result despite variations. Techniques such as Hebbian learning, backpropagation, and others train networks to recognize a standard form for each letter, perhaps with minor variations. Even though the handwritten letter *a* does not always appear the same, nor does it always match a printed *a,* the trained neural network will often output the correct letter. Whereas a conventional computer program requires perfection, neural networks, like brains, make allowances for variety. Computer systems that read handwriting usually rely on neural networks, and are known as intelligent character recognition (ICR), as these systems exhibit a certain degree of "intelligence."

ICR makes life easier for a lot of people who would otherwise have to pore over handwritten documents. For example, Parascript, a company located in Boulder, Colorado, and Lockheed Martin, a company with headquarters in Bethesda, Maryland, developed a system used by the U.S. Postal Service to help process and route letters, even those with handwritten addresses.

These programs are usually neural network simulations. The programs imitate, or simulate, a neural network, although the computers that run the programs have the usual processor and other hardware. Some researchers are working on developing brainlike computer structures—hardware implementations of neural networks—but many researchers and engineers use the simulation programs. The computation is brainlike because of the software's instructions, not because the computer's structure is any different than normal. In such cases, the speed of the computer is an important factor. For jobs in which the input consists of a huge amount of data, a neural network program may be stretched to the limits.

FINDING A NEEDLE IN A HAYSTACK

Suppose someone wants to build a machine that can recognize faces. Humans can perform this task effortlessly, but computers are terrible at it. Software developers have trouble even getting a computer to distinguish a face from the background, let alone to identify and recognize a given face.

Considering the amount of data contained in images, this is not so surprising. A high quality digital camera has a sensor consisting of millions of points, or pixels, that convert light into an electrical signal. The resulting image consists of millions of numbers. Sifting through all that data to find a pattern—a face—buried within it is a time-consuming task for a computer.

But the human eye and brain readily spot contours, angles, and tones that delineate a face. Computers running neural network programs could potentially do the same trick, but are usually bogged down by the massive input. Another difficulty is that neural networks have no teacher, or supervisor, to show them how to do the job. While neuroscientists have made some progress in understanding how the brain processes visual input, no one knows exactly how the brain's neural networks function. Artificial neural networks must learn how to recognize faces themselves, adjusting their synaptic weights based on some principle, such as Hebbian learning, in order to gradually figure out the solution.

Solutions could be more quickly obtained if the input data could be somehow sifted, removing some of the details that are not relevant to the problem at hand. In 2006, University of Toronto researchers Geoffrey Hinton and Ruslan Salakhutdinov developed such a procedure using a neural network called a Boltzmann machine.

A Boltzmann machine is similar to a Hopfield network and uses Hebbian learning, except the activity of the neurons is not rigidly determined by the summed inputs. Boltzmann machines introduce an element of chance, so even when the weighted inputs exceed the threshold, there is a slight chance that the neuron may not fire. Such probabilistic behavior—behavior that is at least partially governed by chance—may not seem desirable, yet in certain cases it provides a much needed amount of flexibility. As an analogy, consider a rigid metal part that needs to fit snugly into a tight space. If the part is cut exactly right then it will fit; otherwise it needs a certain amount of flexibility to adjust to the space. Probabilistic behavior provides this sort of flexibility for neural networks, because without it they tend to get stuck. Neural networks in the brain also exhibit a certain amount of probabilistic behavior. (Boltzmann machines derive their name from 19th-century Austrian physicist Ludwig Boltzmann [1844–1906], who developed mathematical descriptions of the probabilistic behavior used in these devices.)

Winnowing the wheat from the chaff, so to speak, reduces the number of variables and overall data that must be examined. The number of variables is often referred to as the dimension—three-dimensional space, for example, can be described by three variables. The networks used by Hinton and Salakhutdinov decreased the dimension of the problem, allowing them to work more efficiently. Their networks were able to extract salient features of faces as well as classify documents such as news stories based on content. Hinton and Salakhutdinov published their report,

International Neural Network Society

Communication and cooperation among scientists has long been recognized as an essential factor in the growth and development of science. The earliest scientific society, the Royal Society, got its start in England in 1660 when a group of scientists met at Gresham College to discuss ideas. The Royal Society has continued to the present day, fulfilling the vital needs of publishing information and sponsoring meetings. A scientist who does not know what other scientists are doing is likely to waste time in doing experiments that other scientists have already done or pursuing goals that other scientists have already realized are unpromising.

In 1987, Boston University researcher Stephen Grossberg founded the International Neural Network Society. Interest in neural networks had been growing rapidly in the 1980s, due to the work of Grossberg, Boston University mathematician Gail Carpenter, John Hopfield (then at the California Institute of Technology), and others. Conferences

"Reducing the Dimensionality of Data with Neural Networks," in a 2006 issue of *Science.*

As researchers continue to make advances, neural networks will become capable of performing increasingly sophisticated functions. Organizations such as the International Neural Network Society, described in the sidebar above, make it possible for scientists to keep up with the fast pace of research in this field. But some tasks, such as making predictions, remain difficult for artificial neural networks—as well as the real kind.

FORECASTING AND MAKING DECISIONS

In reply to a question about the future of physics, Danish physicist Niels Bohr (1885–1962) once jokingly remarked that predictions are difficult,

play an important role in science by bringing together researchers from around the world into one spot where they can discuss their latest findings. Scientists present results, exchange information, and describe recent experiments. In addition to getting feedback from scientists who are interested in their work, conference attendees also hear about what other scientists are doing. Fruitful collaborations often arise among scientists as they discuss one another's work. About 2,000 researchers attended the first conference sponsored by the International Neural Network Society, which took place in September 1988 in Boston, Massachusetts.

Publishing is also a vital part of science. A publication describes the details of a scientist's efforts, including the specific goals of the research, the results, the methods used to obtain those results, and a discussion of their significance. These publications permit other scientists to study what has already been done, and also establishes credit for the scientists who are the first to publish important discoveries. The International Neural Network Society publishes the journal *Neural Networks* to meet these needs.

especially about the future. Although the complexities of the human brain allow people to plan and forecast the future more often than any other organism, getting it right can be rare. Predicting the course of future events is difficult because there are usually a huge number of unknown factors that will influence the outcome. Accounting for all these factors and forecasting their effects is usually not possible with any great deal of confidence.

Consider the weather, for instance. Newspapers and television news broadcasts always include a weather forecast because the weather affects so many activities and the decisions people must make in advance. While forecasts are usually accurate for the next few hours or a day or two, a forecast made for a week or so in the future almost always misses the mark at least slightly.

Meteorologists—scientists who study the weather—make forecasts by using historical weather patterns along with complicated equations that describe atmospheric activity. Data such as temperature, atmospheric pressure, and wind speed, which are gathered from weather stations and satellites, provide the inputs for sophisticated computer programs. Sometimes the programs assume that future weather will follow a pattern similar to one that has developed earlier under similar conditions of pressure, temperature, and so forth. At other times, the programs plug the data into equations, though these equations are highly simplified models of atmospheric dynamics.

Weather forecasts are hampered by the problem described earlier—large amounts of data—along with uncertainty as to which data are the most relevant for a given prediction. Researchers are investigating a number of different strategies to solve or reduce this problem, including neural networks. Scientists at Johns Hopkins University in Maryland recently developed a neural network that can forecast a more modest but still important type of weather—space weather.

At certain times, the Sun ejects a massive number of particles that travel at high speed through the solar system and cause what are known as interplanetary shock waves. These particles are usually charged and strongly affect Earth's magnetic field, resulting in disturbances in satellite communication as well as an increase in intensity of magnetic phenomena such as aurora borealis (the northern lights, which light up the sky and are caused by charged particles interacting with Earth's magnetic field). Orbiting astronauts can also be adversely affected. No one yet knows how to predict the onset of these events, but once astronomers observe them occurring in the Sun, it is important to predict the time of arrival of the shock wave at Earth's location. Jon Vandegriff and his colleagues at Johns Hopkins University used data gathered from a National Aeronautics and Space Administration (NASA) satellite to train a neural network to predict shock arrival times.

The network consisted of 10 input neurons, one output neuron (which gave the predicted time of arrival), and two hidden layers of fully interconnected neurons. Vandegriff and his colleagues provided the network with data such as particle intensities and the time course of 37 past events, training it to output the correct arrival time. Then the researchers tested the network on 19 other past events, and found that it could forecast the correct arrival time within a reasonable range. Vandegriff and his colleagues published their findings, "Forecasting Space Weather: Predicting Interplanetary Shocks Using Neural Networks," in a 2005 issue of *Advances in Space Research*.

Space weather is not the only phenomenon that neural networks are learning to predict. Moviemakers invest millions of dollars in their product, sometimes for a severely disappointing return at the box office. Ramesh Sharda and Dursun Delen, researchers at Oklahoma State University, trained a perceptron to predict the financial success, or lack thereof, for movies prior to their release. The predictions classified future outcomes in one of nine categories, ranging from "flop" to "blockbuster." Input data included the rating, competition, actors, special effects, genre (type of movie, such as comedy or drama), and the number of theaters showing the picture. The neural network chose the correct category, or the next closest category, 75 percent of the time for previous films. This report, "Predicting Box-Office Success of Motion Pictures with Neural Networks," was published in *Expert Systems with Applications*. As the researchers remarked in the report, their results "prove the continued value of neural networks in addressing difficult prediction problems."

ARTIFICIAL LIFE

The development and use of brainlike technology in computers is a form of biologically inspired computing—researchers and inventors adapt knowledge gained from biology and neuroscience into their computational devices. Continuing in the use of biology for inspiration, computer scientists may go even farther, creating some form of artificial life to go along with a "brain" consisting of artificial neural networks.

A common avenue of research in artificial life is the simulation of biological systems. In this kind of research, computer programs simulate essential aspects of life such as nutrition and genetics, and researchers study the evolution of their simulated life forms. An entire "world" unfolds as the computer program runs. This research can be a lot of fun, and in the 1990s British computer scientist Stephen Grand and his colleagues created a popular game called Creatures. Players of this computer game taught little creatures called Norns how to eat, work, and survive the threats of dangerous creatures such as Grendels. The program simulated biochemical reactions along with genetics, by which the creatures changed and evolved, and a brain made of neural networks. Several versions of Creatures were released, and people continue to play and enjoy it.

Robotics is another area of research that sometimes employs neural networks and other biologically inspired techniques. For example,

Florentin Wörgötter at the University of Göttingen in Germany, along with colleagues at the University of Stirling and the University of Glasgow in Scotland, recently developed a robot capable of walking on two legs over rough or uneven terrain. The robot, called RunBot, has an "eye"—a sensor that detects infrared light—to watch where it is going.

Controlling the motion is a computer that runs programs to simulate neural networks, which consist of simple Hopfield-like neurons. But these networks are different from the artificial neural networks described earlier, since they are organized more like the networks in the brain. The brain generally consists of levels, or tiers, of connected networks, each of which perform a certain function and then pass the results along the chain. RunBot has this sort of hierarchical organization—a structure composed of tiers—and uses Hebbian principles and adjustable synapses so that it can learn how to walk up slopes. (Climbing an incline is an extremely difficult task for robots because it involves a change in stride and center of gravity.) Poramate Manoonpong, Wörgötter, and their colleagues published their paper, "Adaptive, Fast Walking in a Biped Robot

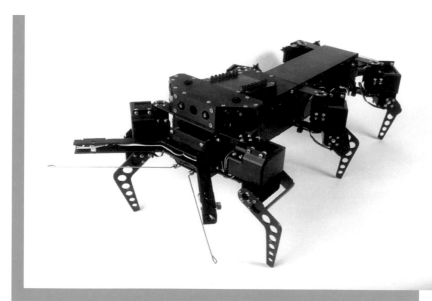

An artificial neural network guides this experimental robotic insect. *(Eurelios/Photo Researchers, Inc.)*

under Neuronal Control and Learning," in 2007 in *Public Library of Science Computational Biology.*

Intelligent behavior, such as learning how to move, is one of the goals of AI. Researchers have made progress but have yet to construct a device even remotely comparable to the human brain. And most of the "brainy" technology described in this chapter is based on computer simulations instead of hardware that emulates brain activity.

But these first steps are necessary. As researchers learn more about neural networks, the computer simulations become more complex and realistic. Implementing these networks in hardware, such as building a machine with a "brain" composed of neuronlike computational elements, is a much more expensive and elaborate procedure. Yet this is already occurring—even as long ago as 1960, when pioneering researcher Frank Rosenblatt built the Mark I Perceptron—and will continue to improve. This gives rise to profound questions about the nature and experience of these machines, including the possibility of some sort of artificial consciousness.

CONCLUSION

Science fiction writer Isaac Asimov (1920–92) began writing a popular series of stories and novels in the 1940s with robots as characters. The robots have an appearance strikingly similar to humans, and are capable of humanlike rationality and behavior, controlled by an instrument Asimov referred to as a "positronic brain." In stories such as "Bicentennial Man," published in 1976 (and made into a movie starring Robin Williams in 1999), Asimov explored the relationship between intelligent robots and humans, as well as the possibility of a robot developing consciousness similar to that of a person.

Although positronic brains are fiction, the neural networks described in this chapter are practical examples of embedding intelligence in machines and computers. Most of these neural network applications have been elaborate programs running on computers. While simulations and computer software can be complex and capable of displaying some measure of intelligence, they are not likely to be conscious. They exist as instructions to a processor, which carries out the instructions in a manner similar to other programs.

But implementing an artificial neural network in its own hardware, such as a positronic brain that controls a robot, raises the same questions that Asimov explored in his science fiction. Although scientists are far from designing the sophisticated robots in Asimov's stories, researchers are making progress. Eduardo Ros Vidal, a researcher at the University of Granada in Spain, and his colleagues are building an artificial brain based on the cerebellum. The cerebellum, or "little brain," is a structure that has many functions in humans, including coordinating movements such as walking or swinging a baseball bat. One of the goals of this research is to incorporate this artificial cerebellum into a robot. Controlled by such a device, the robot will be capable of coordinated movements similar to humans, and will be able to move around and retrieve items in places such as homes, where the robot can act as a helper to handicapped people. This effort is part of a multinational project called SENSOPAC (which stands for sensorimotor structuring of perception and action for emergent cognition), aimed at building cognitive—thinking—robots.

Can an artificial brain ever reach consciousness? Chapter 1 discussed the Turing test, invented by British mathematician Alan Turing, in which a computer is said to be intelligent if its conversational skills are indistinguishable from those of people. No artificial intelligence system has yet to pass this stringent test, and many researchers feel that the Turing test is too difficult for the present state of technology. Another criticism of the Turing test is that it does not adequately define what it means to be "intelligent."

There is even less consensus on what it means to be conscious. No one has devised a test or definition that is universally accepted as embodying the concept of consciousness. Consciousness is an active area of research in neuroscience, and as neuroscientists gain a better understanding of how the brain works, a more complete conception of consciousness may emerge.

Meanwhile, computer scientists, biologists, mathematicians, and physicists are developing artificial neural networks of increasing sophistication and ability. Sometime in the future, robots may display decision-making and forecasting skills comparable to those of a person. If and when this occurs, the question of consciousness will undoubtedly be raised.

At the present time, no one knows if the mental property people call consciousness is exclusively a biological function—and thus requires actual biological tissue—or not. Perhaps some property identical to consciousness can arise from a complicated artificial system embedded in metal and silicon. In any case, people will learn more about the nature of consciousness as artificial neural networks continue to grow in size and complexity, edging closer to behavior that resembles consciousness. The study of artificial neural networks is a frontier of computer science that is not only expanding technology but also providing clues to one of the deepest mysteries of human beings.

CHRONOLOGY

1873 C.E.   Italian scientist Camillo Golgi (1843–1926) develops a staining technique by which neurons can be visualized.

1890s   Spanish anatomist Santiago Ramón y Cajal (1852–1934) uses microscopic analysis and Golgi's technique to study neural networks in the brain.

1897   British physiologist Sir Charles Sherrington (1857–1952) coins the term *synapse.*

1943   Warren McCulloch (1898–1968) and Walter Pitts (1923–69) describe artificial neurons capable of computing logic functions.

1949   Canadian psychology Donald O. Hebb (1904–85) publishes *The Organization of Behavior,* in which he outlines his ideas on learning and synapses.

1950s   Cornell University researcher Frank Rosenblatt (1928–71) develops perceptrons.

1960   Rosenblatt and colleagues build a computer, the Mark I Perceptron, based on perceptrons.

1969   Marvin Minsky and Seymour Papert write *Perceptrons,* a book that criticizes Rosenblatt's research and dampens enthusiasm for artificial neural networks.

1970s   Boston University mathematician Stephen Grossberg develops mathematical principles and descriptions of neural networks.

1974   A thesis written by Harvard University student Paul Werbos contains the first development of backpropagation, an important training technique for neural networks.

1982   California Institute of Technology physicist John Hopfield publishes his work on associative memory and an important type of network known as a Hopfield network.

1986   Much excitement is generated by the publication of *Parallel Distribution Processing,* a two-volume set of books authored by David E. Rumelhart, James L. McClelland, and the PDP Research Group. The books are an accessible treatment of the fundamentals of biologically inspired neural computation, and reach a wide audience of researchers.

1987   Grossberg establishes the International Neural Network Society.

1996   The artificial life game Creatures, which employs neural network simulations, is released.

2000s   Researchers with backgrounds in computer science, mathematics, engineering, physics, and biology begin to develop sophisticated artificial "brains" for applications such as robot control and guidance systems.

2006   University of Toronto researchers Geoffrey Hinton and Ruslan Salakhutdinov publish a method of simplifying data with the use of neural networks.

2007   Florentin Wörgötter and colleagues develop a robot capable of walking on two legs over rough or uneven terrain.

FURTHER RESOURCES
Print and Internet

Beale, R., and T. Jackson. *Neural Computing—An Introduction*. Bristol, U.K.: Institute of Physics Publishing, 1990. Although containing plenty of introductory material, this text also discusses some of the more advanced mathematical principles of the subject and was written for beginning college students. There are chapters on pattern recognition, neurons, perceptrons, and Hopfield networks.

Churchland, Patricia, and Terrence J. Sejnowski. *The Computational Brain*. Cambridge, Mass.: MIT Press, 1992. The idea that the brain functions like a computer is an old one, dating back to the earliest digital computers in the 1940s. But the brain is not a computer, even though it performs "computations" on sensory inputs such as vision, for example, resulting in the perception of visual objects. The authors, a philosopher (Churchland) and a neuroscientist (Sejnowski), discuss how neural networks in the brain function.

George, F. H. "Machines and the Brain." *Science* 127 (30 May 1958): 1,269–1,274. In this early article on neural networks, George describes the mathematical logic and structure of "complex nets."

Hebb, Donald O. *The Organization of Behavior*. Hillsdale, N.J.: Lawrence Erlbaum, 2002. The republication of this 1949 classic allows readers to study and appreciate Hebb's insights into how the brain works. Although subsequent research has dated some of the material, the ideas of this pioneering scientist continue to be relevant.

Hinton, G. E., and R. R. Salakhutdinov. "Reducing the Dimensionality of Data with Neural Networks." *Science* 313 (28 July 2006): 504–507. The researchers found a way to decrease dimensions of certain kinds of problem, and their networks were able to extract salient features of faces as well as classify documents such as news stories based on content.

Hopfield, J. J. "Neural Networks and Physical Systems with Emergent Collective Computational Properties." *Proceedings of the National Academy of Sciences* 79 (1982): 2,554–2,558. Hopfield showed that a certain type of neural network, later called a Hopfield network, has certain computational properties and associative memory.

Manoonpong, Poramate, Tao Geng, Tomas Kulvicius, Bernd Porr, and Florentin Wörgötter. "Adaptive, Fast Walking in a Biped Robot under Neuronal Control and Learning." *Public Library of Science*

Computational Biology (July 2007). Available online. URL: www.ploscompbiol.org/article/info:doi/10.1371/journal.pcbi.0030134. Accessed June 5, 2009. The researchers developed a robot capable of walking on two legs over rough or uneven terrain.

Minsky, Marvin, and Seymour Papert. *Perceptrons: An Introduction to Computational Geometry.* Cambridge, Mass.: Massachusetts Institute of Technology Press, 1969. Minsky and Papert reviewed the potential and the limitations of early neural networks.

Neural Network Solutions. "Introduction to Neural Networks." Available online. URL: www.neuralnetworksolutions.com/nn. Accessed June 5, 2009. Neural Network Solutions, a company located in Cambridge, England, develops neural network technology. This page contains links to introductory material on neural networks, including history and development, artificial neurons, learning algorithms, and applications.

Newell, Allen. "Book Reviews: A Step toward the Understanding of Information Processes." *Science* 165 (22 August 1969): 780–782. Newell reviews the 1969 book *Perceptrons* by Marvin Minsky and Seymour Papert.

Newman, David R., and Patrick H. Corr. "Techniques: The Multi-Layer Perceptron." Available online. URL: intsys.mgt.qub.ac.uk/notes/backprop.html. Accessed June 5, 2009. This Web page presents an overview of the basic ideas of perceptrons, including learning techniques such as backpropagation. Links to JavaScript models are included.

Parascript, LLC. "The History of ICR & OCR." Available online. URL: www.parascript.com/company2/tech_overview.cfm. Accessed June 5, 2009. Parascript develops software for analyzing visual information. This Web resource describes optical character recognition, intelligent character recognition, and the differences between the two.

Sarle, Warren. "Neural Network Frequently Asked Questions." Available online. URL: faqs.org/faqs/ai-faq/neural-nets/part1. Accessed June 5, 2009. In the 1990s, when the Internet started to become popular, newsgroups formed in which participants engaged in online discussions by posting their ideas or questions. To avoid cluttering the newsgroup with questions that are asked repeatedly, a list of frequently asked questions (FAQ) was made available. This Web resource contains the first part of the FAQ for a newsgroup devoted to neural networks and links to the other parts. Although this document is no longer updated, there is plenty of valuable material for a reader interested in the basic concepts of neural networks.

Sharda, Ramesh, and Dursun Delen. "Predicting Box-Office Success of Motion Pictures with Neural Networks." *Expert Systems with Applications* 30 (2006): 243–254. The researchers trained a perceptron to predict the financial success, or lack thereof, for movies prior to their release.

Stergious, Christos, and Dimitrios Siganos. "Neural Networks." Available online. URL: www.doc.ic.ac.uk/~nd/surprise_96/journal/vol4/cs11/report.html. Accessed June 5, 2009. This page contains a comprehensive, well-illustrated article about artificial neural networks. Discussions include history, similarities with biological neural networks, training, structures, and applications.

Vandegriff, Jon, Kiri Wagstaf, George Ho, and Janice Plauger. "Forecasting Space Weather: Predicting Interplanetary Shocks Using Neural Networks." *Advances in Space Research* 36 (2005): 2,323–2,327. The researchers used data gathered from a NASA satellite to train a neural network to predict shock arrival times.

Web Sites

Cyberlife Research. Available online. URL: www.cyberlife-research.com. Accessed June 5, 2009. Computer scientist Stephen Grand formed his own company, Cyberlife Research, with the aim of developing intelligent artificial life. Grand, the cocreator of the artificial life game Creatures, provides a lot of information on this and other projects at this Web site.

University of Toronto Computer Science Department. Geoffrey E. Hinton Web page. Available online. URL: www.cs.toronto.edu/~hinton. Accessed June 5, 2009. Hinton, a professor at the University of Toronto, is an expert on neural networks. On this Web page are links to some of his publications, tutorials, and videos.

COMPUTATIONAL COMPLEXITY AND SOLVING PROBLEMS EFFICIENTLY

Complexity theory in computer science refers to the study of the amount of computer time and resources required to solve problems. Finding a method to solve a problem is not necessarily the end of the job—even if a certain method works, if may not provide an answer in a reasonable amount of time. In computer terms, a problem is a question to be resolved or decided by some kind of computation. One of the most critical frontiers of computer science involves determining which problems can be solved in a reasonable amount of time. Consider the following example.

Suppose that Steve is a sales representative for a newly founded company. To increase the number of orders, he decides to make a tour through several cities and visit potential customers. But as an employee of a new company, which typically has limited resources, he also needs to be economical. The sales representative wants to visit the cities on his schedule in an order that minimizes the length of the trip. In other words, Steve wants to leave home and visit city A first, B next, and so on until he arrives back home, in an order that covers the least amount of distance. If he is driving, this arrangement would be the least expensive in general because it would require less gasoline than any other route. The shortest route is usually not at all obvious if the cities are spread out over a large area. An example appears in the sidebar on page 72.

There may be other factors to consider, such as the weather and the condition of the roads, and if the sales representative flies or rides a train instead of driving, he will concentrate on finding the cheapest tickets. Practical problems involving some type of minimization or optimization arise often in business, engineering, and science. The problem described above is called the traveling salesperson problem (TSP).

Any given instance of this problem can be solved by calculating all the possibilities, then choosing the right one. Computer scientists sometimes call this the "brute force" approach because it tackles the problem in a direct and unsophisticated manner. Such an approach will always be successful if the problem solver has sufficient time. For example, a TSP with four cities, A, B, C, and D, has 24 possible routes: A → B → C → D, B → C → A → D, and 22 others. But half of these routes will have identical distances because they will be the reverse of another route—for instance, D → C → B → A will cover the same distance as the first route listed above because it is the exact reverse and the distance from A → B is the same as that for B → A, and so on. If Steve calculates all 12 possible distances, he can find the minimum.

The difficulty with this approach becomes apparent as the number of cities increases. If Steve must visit 10 cities, the number of routes is 3,628,800 (1,814,400 distances)—quite a lot of calculating! Finding a minimum distance is extremely time-consuming using the brute force method for this many calculations. For 60 cities, the number of routes is comparable to the estimated number of particles in the entire universe (about 10^{80}). The rise in the number of calculations is so steep that even with the increasing speed of computers and the use of advanced technology such as neural networks, people still cannot solve these problems by brute force in a reasonable amount of time.

A more sophisticated approach is needed for problems that are computationally complex. Research on computational complexity impacts everyone, from sales representatives planning trips to consumers who make a purchase over the Internet with a credit card, and is one of the most active frontiers of computer science. But the topic involves some difficult concepts, and can be challenging for researchers as well as students. This chapter provides a gentle introduction that presents the basic ideas and issues that drive much of the current research in this field.

INTRODUCTION

The rise of digital computers, chronicled in chapter 1 of this book, led to an increasing reliance on automated computation in society. By the 1960s, most businesses and universities were using computers for a variety of purposes, such as performing lengthy scientific and engineering calculations or maintaining sales records. Most computers are general-purpose machines capable of calculating whatever needs to be calculated, but such machines need an accurate set of instructions to do so. If a computer had a specific built-in purpose, the instructions could be wired into the hardware, but in this case the machine would be useful only for this specific job. General purpose machines are versatile, although the price the user pays for this advantage is the need to write the instructions or programs. This means that computer users must identify the algorithm—a procedure or set of instructions—to solve the problem at hand.

The widespread use of computers by the 1960s resulted in a large number of users writing their own programs to solve problems. In many cases, there is more than one way to solve a problem, just as there exist a number of different ways of traveling from one point to another. But not all algorithms are created equal. Programmers began to discover that some algorithms were more efficient than others, performing the same task in much less time or with fewer computer resources, such as memory space and disk storage, than other algorithms. Similar to traveling, where the shortest distance between two points is a straight line, some algorithms took the expressway and some meandered all over the place, requiring an excessive amount of time and resources—and taxing users' patience as they waited for a program that seemed to take forever to finish.

Although some programs were obviously slower than others, no explanation for this phenomenon was readily apparent. Computers were new territory. But as scientists and mathematicians began to study and classify algorithms, some answers started to emerge. In 1965, Cornell University researcher Juris Hartmanis and his colleague, Richard Stearns, quantified the complexity of algorithms by considering the number of steps a simple computer required to perform them. This work "introduced a new field and gave it its name," the Computer Science Department at Cornell University noted in October 2005, celebrating the achievements of its researchers.

Prior to the 1960s, mathematicians and computer science pioneers had focused on the question of computability rather than complexity.

Complexity involves many choices, as in this depiction of a large number of overlapping pathways *(DSGpro/iStockphoto)*

Computability is the study of what kind of problems can be solved in principle, regardless of the time and resources necessary to do so. Not all problems have solutions—as a simple example, consider the problem of finding the last digit in the decimal representation of π (the Greek letter pi), the ratio of a circle's circumference to its diameter. This mathematical constant is approximately 3.14159, but its precise value is an irrational number, meaning that the decimal representation never ends or repeats.

Mathematicians and early computer scientists were especially interested in systems by which any theorem could be proved and all problems solved. The study of complexity is different because it focuses on the issue of how efficiently the solutions to a problem can be, assuming a solution exists.

As computer expert Grace Hopper and others discovered in the 1940s, computers and computer programs often have "bugs" causing errors or slowdowns. Although bugs must be avoided from a practical standpoint, they are excluded in the study of algorithmic complexity, which assumes the algorithm is or will be correctly implemented. Errors may arise, but these mistakes will, or should be, discovered and corrected, so bugs and mistakes play little role in the formal study of algorithms.

Scientists who study computational complexity also do not generally consider the details of the computer that will run the programmed algorithm. (But quantum computers, discussed in chapter 1, will present special problems if they are ever developed, as described below.) Computational complexity is an abstract subject and involves a great deal of advanced or abstract mathematics. It is not an easy subject to grasp, and this chapter will present only some of its elementary aspects. But even with the simplifications, the richness and excitement of algorithms and computational complexity are apparent.

ALGORITHMS

An algorithm can be written as a series of steps. Computers only do what they are told, so the algorithm must not omit any important steps, otherwise the problem will not be solved. But adding unnecessary steps is unwise because this will increase the program's run time.

Computers operate with a binary language called machine language—words or strings of 1s and 0s—that is difficult and cumbersome for programmers to use. Although all programs eventually must be translated into machine language, most programmers write programs in languages such as BASIC or C, which contain a set of commands and operations that humans can more readily understand. Programs called compilers or interpreters then translate these instructions into machine language.

Programmers often begin with an outline or a sequence of steps the algorithm is to accomplish. For example, consider the problem of finding the largest number in a list. The steps may be written as follows.

1. Input the list of numbers.
2. Store the first number in a location called Maximum.
3. Store the next number in a location called Next and compare the value of Maximum with that of Next.
4. If Maximum's value equals or exceeds that of Next, discard the number in Next. If not, replace the value of Maximum with Next's value.

5. Repeat steps 3–5 until the end of the list is reached.
6. Output Maximum.

More complicated problems involve a lot more steps. Deep Blue, the IBM computer that beat reigning champion Garry Kasparov in chess in 1997, ran a considerably longer program.

Finding the maximum number in a list is a simple problem with a simple solution, but even simple solutions may take a long time. The TSP is an excellent example. Because it is a practical problem as well as representative of a large class of interesting problems in computer science, it has been much studied. Writing in the February 15, 1991, issue of *Science,* researchers Donald L. Miller, at E. I. du Pont de Nemours and Company, in Delaware, and Joseph F. Pekny, at Purdue University, in Indiana, observed, "Few mathematical problems have commanded as much attention as the traveling salesman problem." The sidebar on page 72 provides a sample TSP calculation.

A brute force approach to solving a TSP would calculate each possible distance. A more thrifty approach is to use *heuristics*—shortcuts to speed up the computation, such as the author's assumption, as mentioned in the sidebar, that the shortest route would not contain the longest intercity distance. Shin Lin and Brian Kernighan published a heuristic in 1973, "An Effective Heuristic Algorithm for the Traveling-Salesman Problem," in *Operations Research,* that uses a complicated procedure. But these procedures are not guaranteed to find the optimal solution. Instead, the goal is to find a good solution in a reasonable amount of time.

SCALABILITY—INCREASING THE SIZE OF THE PROBLEM

Heuristics can shave some time off a lengthy computation, but they cannot generally find the best solution. And when number of possibilities rises steeply, as it does in TSP, the solution may not be all that great. The ability of an algorithm to perform well as the amount of data rises is called *scalability*. An increase in data usually means an increase in the number of calculations, which results in a longer run time. To scale well, an algorithm's run time should not rise prohibitively with an increase in data.

Consider an algorithm that searches for every instance of a certain word in a document. The simplest algorithm would examine one word,

Chess match between Garry Kasparov and IBM's Deep Blue *(Stan Honda/AFP/Getty Images)*

see if there is a match, and then move on. If the size of the document is doubled, then the number of words will be about twice as large, and the search algorithm will take twice as long to search, all other things being equal. A document four times as large would take four times as long. This increase is linear—the increase in run time is proportional to the increase in data.

Now consider the brute force approach to solving a TSP. The number of possible routes rises at an astonishing rate with even a slight increase in the number of cities, which is a much worse situation than in the search algorithm. If the input of the search algorithm goes from four words to eight, then the time doubles; for brute force TSP calculations, an increase from four cities to eight increases the time by a factor of 1,680. The search algorithm scales at a tolerable level, but the brute force TSP is a disaster.

Advances in computer technology discussed in chapter 1 have resulted in blazingly fast computers, yet even with these speedy machines, algorithm scalability is important. Linear increases, such as the simple search algorithm displays, do not present much of a problem even with relatively large data sets, but this kind of tame behavior is the exception rather than the rule. The TSP is a notoriously difficult case, but there are many other problems that scale poorly. Even with heuristics and approximations, solving these problems when the data set is large takes an extraordinary amount of time, if it is possible at all. To the chagrin of scientists and engineers, many of the more interesting and important problems in science and engineering belong in this category.

There is another practical concern besides run time. Computers hold instructions and data in memory as the program runs. Although some computers have a huge amount of memory, it is still a limited resource that must be conserved. Scientists and engineers who share the use of the most advanced supercomputers are allotted only a certain amount of memory, and users who pay a supercomputer facility to use one of their machines are often charged for the memory as well as the time their programs consume.

Memory can also be an important factor in the run time of a program. Computer memory consists of fast devices such as random access memory (RAM) circuits that hold millions or billions of chunks of data known as bytes, so retrieving a piece of data or loading the next instruction takes little time. But if a program exceeds the amount of available memory, the computer must either stop or, in most cases, will use space on the hard disk to store the excess. Disk operations are extremely slow when compared to memory, so this procedure consumes a great deal of time. A personal computer, for example, should have ample memory, otherwise large programs such as video games will be uselessly slow even if the processor is the fastest one available.

The memory versus speed trade-off was more imperative back in the 1980s, when memory was measured in thousands of bytes rather than today's billions. Although memory is still an important factor in certain applications, the main issue in general is time. When a programmer examines an algorithm's efficiency, he or she is usually concerned with its speed.

An algorithm's complexity can be measured by its run time—this is called time complexity—or the amount of memory resources it uses, which is known as space complexity. But speed, like the amount of memory, depends on the computer—some computers are faster than others. A measure of run time will apply only to a certain machine rather than to all of them.

To avoid the need to analyze an algorithm for each kind of computer, a general measure, such as the number of steps required by the algorithm, can be used. The number of steps does not usually change very much from

The Traveling Salesperson (or Author) Problem

Suppose the author embarks on a four-city tour to promote his books and meet thousands of readers (or at least one or two). The figure illustrates a rough map of the author's home, marked with an H, and the four destinations, labeled A, B, C, and D. As shown in the map, the distances between locations are as follows: A to B (and B to A) = 5 units, A to C = 13, A to D = 6.7, B to C = 10.8, B to D = 10, and C to D = 11.7. The distance from the author's home to A is 4.3 units, to B is 5.6, to C is 9.1, and to D is 4.6. The author will leave from home and visit each city in turn, then return home. What is the order that minimizes the amount of distance the author must travel?

A simple algorithm to solve this problem is to compute the distance of all possible routes. There are 24 routes in this TSP problem, although as mentioned earlier, 12 of them are simply the reverse of another. To see that there are 24 possible routes (including reversals), consider the first visit—there are four choices to make here. Three remain, so the next decision involves three choices, then two, and then one. The number of routes equals the product of the number of choices, which in this case is $4 \times 3 \times 2 \times 1 = 24$.

When the author set up this problem, he anticipated that the shortest path would avoid the longest distance, which is the 13 units between A to C. In other words, the author expected that he would not want to go to A directly from C, or C directly from A. This proved to be correct. He also believed that the shortest path would probably avoid the next longest distance, which is the 11.7 units between C and D. This hypothesis was also correct, but just barely. The shortest path is H → C → B → A → D → H, which is 36.2 units. (The

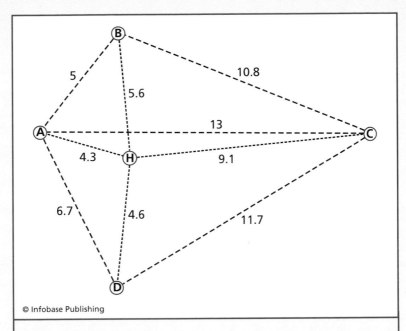

© Infobase Publishing

The author needs to leave home, H, and visit each of the cities A, B, C, and D via the shortest possible route. Values are given for each intercity distance as "units."

reverse order is also 36.2 units.) The next shortest route, only 0.2 units longer, is H → A → B → C → D → H (and the reverse), and it requires a trip between C and D. Coming in at 44.6 units is the longest route, H → A → C → D → B → H (and the reverse).

Other factors may complicate this problem, such as a desire to return home on one leg of the trip to rest. But the simplest scenario has a solution that in this case was relatively painless to calculate. Yet even with this simplicity, the shortest route was not obvious, at least to the author, before making the calculations.

machine to machine, but it does depend on the size of the problem, as well as the specific outcome. For example, a search algorithm may end early if the target is found, truncating the rest of the search. In order to be as general as possible, an analysis of an algorithm usually considers the worst-case scenario, which requires the most steps. For the example of a search algorithm that quits when the target is found, the worst case is that the target is in the last piece of data, or is not found at all.

Since the number of steps an algorithm requires to solve a worst-case scenario depends on the size of the problem, computer scientists describe the number of steps in terms of n, a variable representing the size of the input data. A linear increase, such as the search algorithm described earlier, is represented as n. If the number of steps increases with the square of the input, the value is n^2.

Determining the number of steps required by an algorithm is complicated. Most of these calculations are rough approximations, sometimes known as order of magnitude approximations because the size is estimated to either the nearest order of magnitude—a factor of 10—or an even higher order, such as the size of an exponent. In these rough approximations, a value of 20 times n^2 ($20n^2$) would simply be estimated n^2; the value n^2 would only be distinguished from a sizeable difference, say, n^3. This order of magnitude notation is sometimes called big O, and is written, for example, as $O(n)$ for the linear case, $O(n^2)$ for n^2, and so forth. Note that when n is large, $\log n < n < n^2 < n^3 < 2^n$. (Log n is the logarithm of n—the power to which the base number, usually 10 unless otherwise specified, must be raised in order to produce n. For example, $\log 100 = 2$, because $10^2 = 100$.)

Functions that contain n or n raised to a power, such as n^2, are known as *polynomial* functions. If a number is raised to the nth power, such as 2^n, it increases exponentially as n increases, and the function is called an *exponential* function.

CLASSES OF COMPLEXITY

Algorithms offer solutions to problems, and the scalability of an algorithm determines the feasibility of solving large instances of a problem. The important question of what kind of problem can be solved in a reasonable amount of time comes down to the algorithms and their scalability.

Problems can be categorized into classes based on their complexity, as measured by the algorithm that solves the problem. The problem of searching for a word in a document is a simple problem because there is a simple algorithm to solve it. The TSP, in contrast, is an extremely difficult computational problem. A complexity class is a set of problems that can be solved by algorithms of a certain complexity; that is, algorithms requiring a certain number of steps.

Determining the number of steps an algorithm takes can be a difficult task. Classifying problems is even more difficult, because there is usually more than one algorithm to do the job. The number of steps of these algorithms generally vary, with more efficient algorithms doing the job with fewer instructions. When studying the complexity of a problem, computer scientists consider only the most efficient algorithm. For example, if a programmer discovers an algorithm that scales as $\log n$ instead of n^2, the problem has a complexity of $\log n$.

Proving that an algorithm is the fastest possible solution to a given problem is also an arduous task. In many problems, computer scientists are not certain that the most efficient algorithm known is really the best. But the most important distinction is made between two strikingly different magnitudes—polynomial and exponential. In the 1960s, Jack Edmonds, then at a government agency known as the National Bureau of Standards, was one of the first researchers to study the differences between polynomial and exponential complexities. (The National Bureau of Standards is now called the National Institute of Standards and Technology, as discussed in the sidebar in chapter 1.) Most (though not all) of the problems with polynomial complexity prove to be solvable in a reasonable amount of time—these are doable problems. Problems with exponential complexity are not feasible, except for small amounts of data or in other special cases.

This distinction is critical because it separates what computer science can expect to achieve from what it cannot. Even as computers and computer technology advance in size and speed, problems of exponential complexity will remain out of reach unless some kind of scientific breakthrough occurs. Quantum computers might be one such breakthrough, but as far as foreseeable developments are concerned, polynomial complexity is generally the limit.

In studying computational complexity, computer scientists have focused much of their attention on two complexity classes, P and NP.

The complexity class known as P is the set of problems that can be solved with polynomial run-time algorithms. Simple search problems are in this set. The other class is tougher to describe. This class, NP, can be thought of as the set of "reasonable" problems—problems that have an easily verified solution. The method of obtaining this solution is unimportant, as long as the solution can be verified by some kind of algorithm having a polynomial complexity. The following sidebar provides more details.

P is a subset of NP, as described in the sidebar. It might seem likely that any problem imaginable is in NP, for verifying a solution would not appear to be especially difficult once the solution has been found. But computer science theorists have explored a number of complicated problems, some of which are fundamental to the field of computability, which as mentioned earlier is the study of what problems can or cannot ever be computed. Problems outside of NP do exist, although they will not be discussed here.

One of the most important questions in the field of computational complexity is the relationship between P and NP. P is a subset of NP, but is NP a subset of P? If so, then NP = P, or in other words, the two sets are identical. This would mean that all NP problems can be solved by algorithms of polynomial complexity, even the ones for which no such algorithm is presently known to exist.

The consequences of the discovery that these two classes are identical would be huge. Many important problems are in NP, and are solvable today only with algorithms of exponential complexity. For instance, TSP, which can only be approximated by such algorithms, let alone solved, is in NP. If the mathematical statement P = NP is true, it would imply that problems such as TSP can be solved by efficient, polynomial algorithms, waiting to be discovered. This is why many computational complexity theorists have applied their mathematical and logical skills to prove or disprove the equivalence of P and NP.

NP-COMPLETE PROBLEMS

Much of the research on the relation of P and NP involves considerably advanced mathematics. Yet the focal point of this research has boiled down to a special class of problems in NP known as NP-complete. These are some of the thorniest problems in NP, and if researchers can find

Stock traders often employ complicated algorithms to help them decide when to buy or sell. *(Flying Colours, Ltd./Getty Images)*

an efficient algorithm—that is, an algorithm of polynomial complexity—to solve just one of these problems, then all of them have efficient algorithms. In other words, if one problem in NP-complete is also in P, then they all are.

An important tool in defining the set of NP-complete problems is the reduction of one problem to another. Researchers use mathematical techniques known as transformations to change one problem into another. This procedure is similar to translating a book from one language, such as Spanish, to a book written in another language, such as English, without changing any of the important concepts. If the mathematical transformation of one problem to another can be accomplished in reasonable (polynomial) time, then the problems are essentially equal in difficulty. If a certain algorithm can solve one of the problems, then it can solve the other problem after performing a suitable transformation, which does not require too much time.

NP-complete problems are the "hardest" problems in NP in the sense that all NP problems reduce to them via polynomial transformations. This is how the set of NP-complete problems are defined—an NP-complete

NP and P

The importance of polynomial complexity stems from its solvability in a practical amount of time. Problems having polynomial complexity are sometimes referred to as easy, which can be misleading since the algorithm that solves the problem might be extremely complicated and require a run time that is beyond most people's patience. Yet the solution is obtainable, at least in principle, in a reasonable amount of time, which is the important point. The complexity class P is the set of these problems, and the letter *P* stands for polynomial.

NP is the set of problems whose solution can be verified by algorithms of polynomial complexity. This is not the same definition as given for P—problems in the P class have solutions of polynomial complexity, rather than just a verification of the solution. The solution to a problem in NP, once it is known, can be checked in a reasonable amount of time. P is a subset of NP—all problems in P are also in NP—since finding a correct solution to a problem is verified by the act of doing so.

There is another, equivalent definition of NP, which is the source of the class's name. NP stands for nondeterministic polynomial. A nondeterministic algorithm can follow multiple paths at once. Suppose an algorithm reaches a point where it must branch, such as an instruction to take one action or another. These instructions are common in programming, such as "if" statements—if A is true, do this, otherwise do that. A deterministic algorithm can only follow one path, but a nondeterministic algorithm can follow the options simultaneously. All algorithms that actually exist today are deterministic; nondeterministic algorithms or machines are fictional, but they are convenient for computer scientists to use in classifying problems and in theoretical studies. NP is the class of problems that can be solved in a reasonable amount of time with a nondeterministic algorithm.

problem is in the set NP, and all other NP problems can be reduced to it. Since the reduction is "easy," the definition means that no problem in NP is more difficult to solve. (The term *complete* generally refers to a problem that is as hard as or harder than any other in its set, as NP-complete problems are for NP.) The figure diagrams the sets P, NP, and NP-complete.

Harvard University student and later University of Toronto computer scientist Stephen A. Cook introduced the concept of NP-complete in 1971, and showed that a problem called satisfiability is a member. (Satisfiability involves proving the truth of certain formulas.) Thanks to Cook's efforts, subsequent researchers can show an NP problem belongs in NP-complete by showing it can be transformed into this problem, or into some other NP-complete problem that has already been shown to transform to this problem. The transformations are like links in a chain, relating these problems to each other.

Thousands of problems have been shown to be NP-complete. The list includes TSP, which was one of the earliest problems proved to belong to this set. Many of these problems are similar to TSP in that they are involved in optimization—finding the best or optimal solution in a situation. These problems are interesting because they are so hard, but they are also important from an economic standpoint. Solutions to these problems would save time and money, but because of the algorithmic complexity, only approximations, with varying degrees of success, have been found.

NP-complete problems crop up in many fields of technology, including the design of large-scale computer networks and other connected systems, data compression (reducing the amount of data with no or little loss of meaning, which is important in storing large amounts of data such as motion pictures on a disk), *database* searches, and event scheduling. These problems also occur in many branches of computer science and mathematics, as well as in games such as Tetris (a video game in which the player stacks falling blocks) and sudoku (in which the player completes a grid).

Although special instances of some of these problems can be solved easily, no one has ever found a general solution to any NP-complete problem that is polynomial. All of the known algorithms are exponential at best.

Yet if an intrepid researcher finds an algorithm of polynomial complexity for even one NP-complete problem, every member of the set must have one as well. And if this is the case, then all of NP must have polynomial complexity—NP-complete problems are by definition the hardest in NP. This would mean that NP = P.

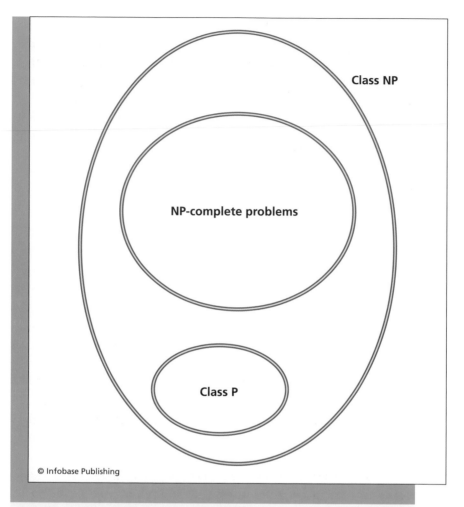

Class NP

NP-complete problems

Class P

© Infobase Publishing

Set NP contains set P and NP-complete.

Considering the importance of these problems, no lack of effort has been spared to find efficient algorithms for these problems. The lack of success thus far suggests, though it does not prove, that such algorithms do not exist. And additional NP-complete problems are continually being discovered. Computer scientist A. K. Dewdney, in his 1993 book *The New Turing Omnibus,* wrote, "The existence of so many NP-complete problems has resulted in a certain attitude of pessimism. Each new NP-complete problem merely reinforces the idea of how hopeless it is to hunt for a polynomial-time algorithm for an NP-complete problem."

A researcher could conclusively demonstrate that NP ≠ P by proving that at least one NP-complete problem cannot be in P. If so, then NP cannot be the same as P. But to date no one has found such a mathematical proof. Even so, the lack of success in finding polynomial-time algorithms for NP-complete problems has created a certain amount of pessimism, as Dewdney notes. In the 2006 book *Algorithms,* Sanjoy Dasgupta and his colleagues wrote, "Are there search problems that cannot be solved in polynomial time? In other words, is P ≠ NP? Most algorithms researchers think so."

But the question remains open. If NP = P, then a lot of hard problems are actually easy. If NP ≠ P, then there exists a nonempty set of problems whose solution is reasonable to check, but no easy algorithm to find the solution will ever be discovered. Emphasizing the importance of this question, the Clay Mathematics Institute will award a million-dollar prize to the first person to find and prove the answer. As described in the following sidebar, this prize is one of seven identified by the institute as "Millennium Prize Problems"—critical problems that have so far resisted all attempts to solve them.

The interrelatedness of NP-complete problems means that if a person could solve TSP efficiently, he or she could also in principle design complicated networks, compress data, schedule airline traffic, and perform many other important tasks a lot more efficiently than they are now. But many researchers are skeptical, and once a specific problem has been shown to be NP-complete, these researchers assume the problem will never be easy to solve. The same skeptical viewpoint holds for problems known to be in NP but not presently thought to be in P—these are hard problems (although not necessarily the hardest) with no efficient algorithm yet found for them. This skepticism has influenced the development of certain high-profile applications, including the security systems protecting personal information traveling over the Internet.

INTERNET SECURITY

To keep secrets safe, governments and militaries have long used some form of *encryption,* in which a message is scrambled so that an enemy will not be able to interpret it if the message is intercepted. The intended recipients can unscramble the message because they know the encryption scheme, but without this knowledge the message appears random. Encryption will work if the amount of time required to unscramble a message without knowing the encryption scheme is an unreasonably long time, or in other words, if there is no efficient algorithm to break the encryption. The methods of encryption, along with attempts to break these methods, are an important frontier of computer science that will be discussed in chapter 4.

Keeping information safe from prying eyes is also vital on the Internet. Each day millions of dollars in transactions occur via this vast computer network, exposing credit card numbers, bank account information, Social Security numbers, and other personal information. The Internet consists of a huge number of computers that communicate with one another, and information passed from one computer to another can be intercepted with ease. Thieves who acquire someone's personal information can assume the person's identity—an act known as identity theft—and make purchases, open credit accounts, and engage in other transactions that will be charged to the victim. The Federal Trade Commission, a United States government agency regulating trade and commerce, estimates that each year about 9 million Americans become victims of identity theft.

Encryption must present a problem of unreasonable complexity for snoopers and thieves. A common type of encryption used on the Internet incorporates a mathematical process called *factorization*—factoring a number means breaking it down into a set of smaller numbers whose product equals the original number. The prime factorization of a number is a set of primes—numbers divisible only by 1 and the number itself— whose product equals the original number. For instance, 5 is a prime because the only numbers that divide it are 1 and 5. The prime factorization of 20 is $2 \times 2 \times 5$, as illustrated in the figure on page 84. This factorization is unique because according to the mathematical principle known as the fundamental theorem of arithmetic, every positive integer has only one prime factorization. Although 20 also equals 4×5, 4 is not prime because 2 is a divisor.

Although the fundamental theorem of arithmetic guarantees that each positive integer has only one unique factorization, no one knows of a simple algorithm to find it. Prime factorization of a small number such as 20 or even 1,000 is not difficult, but for very large numbers the problem is time-consuming—in other words, the problem has a significant computational complexity. Internet encryption methods make use of the prime factors of a huge number, say N, to encrypt a message. Unintended recipients may intercept the message but they will not be able to interpret it without factoring N.

Clay Mathematics Institute and Millennium Prize Problems

On August 8, 1900, at a conference in Paris, France, influential German mathematician David Hilbert (1862–1943) presented a list of then-unresolved problems he considered most worthy of attention. These problems, of which there were 23, became the targets of a great deal of mathematical research, and a number of them have been solved, spurring advances in a broad variety of mathematical topics. One hundred years later, on May 24, 2000, at another conference in Paris, members of the Clay Mathematics Institute decided to ask mathematicians to submit ideas for a new list of problems. The hope was to energize research again, and this time, some extra motivation was added—a bounty of 1 million U.S. dollars for each problem.

Founded in 1998 by Landon T. Clay and his wife, Lavinia D. Clay, the Clay Mathematics Institute is a nonprofit foundation based in Cambridge, Massachusetts. The institute encourages mathematical research, recognizes the achievements of mathematicians, and supports mathematical education. One of the ways the institute pursues these goals is by offering prizes for the solution of important and as yet unsolved problems. The list solicited in 2000 came to be known as the Millennium Prize Problems. Seven problems made the list, including P versus NP. The solution to each problem is worth a million dollars.

An interesting aspect of the P versus NP question is that if P = NP, a great many problems will turn out to have "easy" solutions. This could include some or even all of the other problems on the Millennium Prize list!

Factorization of large numbers is an extremely difficult problem, belonging in the set NP. Yet it does not appear to be an NP-complete problem. This raises few concerns unless factorization is also in P, the set of

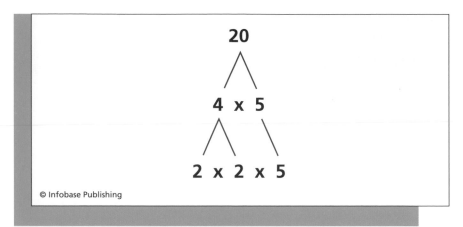

© Infobase Publishing

The number 20 factors into 2 × 2 × 5.

problems solved by algorithms of polynomial complexity. Nobody knows of an "easy" algorithm for this problem, and encryption experts assume one does not exist, so factorization seems to be in NP but not in P.

But the field of computation and computational complexity continues to advance. In 2002, Manindra Agrawal, Neeraj Kayal, and Nitin Saxena of the Indian Institute of Technology Kanpur in India found an algorithm of polynomial complexity that can determine if a given number is prime. This is not the same problem as prime factorization, yet the discovery was surprising because it means that determining if a number is prime is a problem in P. And without proof, computer scientists cannot be absolutely certain that prime factorization is also not in P.

The discovery of an efficient algorithm for prime factorization would force drastic changes in Internet technology. And if a person ever found such an algorithm but did not tell anyone else, he or she would be able to learn an enormous number of secrets.

CONCLUSION

Researchers in the field of computational complexity have made much progress since the 1960s, when the widespread usage of computers motivated a search for efficient algorithms. Scientists have identified important classes such as P and NP, and found plenty of examples of NP-complete problems, the hardest ones. But there is much left undone. No one is yet certain if NP = P, which would imply that there exist efficient algorithms for NP problems, or if—as most computer scientists suspect—there are problems in NP that are not generally solvable with algorithms of polynomial complexity. A lot rides on this question, including the integrity of many forms of encryption and the security of the Internet.

Another undecided issue concerns the development of more advanced forms of computation. Algorithms of exponential complexity would seem beyond the bounds of conventional computers, at least for large data sets, but what about a computer that can take shortcuts? As described in chapter 1, quantum computation is a process that scientists hypothesize could employ principles of advanced physics to accelerate the rate of calculations. Computers based on quantum mechanics, which governs the behavior of small particles such as atoms, would be much more powerful than conventional computers because they would be able to make many calculations at once.

Quantum computers do not exist at the present time, but there is no law of physics that precludes them, and researchers have already taken the first steps to building one. (See chapter 1.) Peter Shor, while working in 1994 at Bell Laboratories, in New Jersey, discovered an algorithm capable of prime factorization in polynomial time—if running on a quantum computer. If quantum computers are developed, the security of the Internet will be compromised until a new encryption strategy is adopted.

Researchers are not sure if quantum computers could solve NP-complete problems in polynomial time. (Recall that factorization is not believed to be an NP-complete problem.) Perhaps encryption standards should be shifted to an NP-complete problem, although this may not eliminate the danger.

The question of NP versus P is unanswered, yet scientists continue to chip away at a solution, sometimes by determining which approaches cannot possibly work. Steven Rudich, at Carnegie Mellon University in Pittsburgh, Pennsylvania, and Alexander A. Razborov, at the Steklov Mathematical Institute in Moscow, Russia, won the prestigious Gödel Prize in 2007 for showing a broad class of mathematical proofs that cannot solve the issue. The Gödel Prize, named after Austrian mathematician Kurt Gödel (1906–78), is awarded annually by two groups—the European Association for Theoretical Computer Science and the Association for Computing Machinery (ACM)—for outstanding work in theoretical computer science. Rudich and Razborov won the prize for work they presented in 1994 at the Twenty-Sixth Annual ACM Symposium on Theory of Computing, in Montreal, Canada.

The discovery of Rudich and Razborov does not rule out a mathematical proof for the question of NP versus P, but it does eliminate a lot of options for finding one. Although many people believe that P does not equal NP because programmers have not found efficient solutions to the hardest problems in NP despite years of searching, the question must remain open until a proof is found.

Perhaps the question will remain open for a long time. But even so, confidence that there are hard problems with no "easy" solutions increases as the years go by. When a researcher discovers that a problem he or she is working on is an NP-complete problem, this is useful knowledge because the researcher realizes that no efficient solution is yet known—and unless NP = P, an efficient solution does not exist. Without this knowledge, a lot of time could be wasted in a fruitless search.

But there is a caveat: No one can be absolutely certain that P does not equal NP. The constraints on what computers and algorithms can accomplish is a vital and undecided issue at the frontier of computer science.

CHRONOLOGY

1930s	British mathematician Alan Turing (1912–54) and other researchers begin developing models and theories of computation.
	Although problems similar to the traveling salesperson problem had been studied earlier, Austrian mathematician Karl Menger begins investigating the general properties of this problem.
1960s	As computers become increasingly prevalent in technology, computer scientists study the efficiency of algorithms.
1965	Cornell University researcher Juris Hartmanis and his colleague Richard Stearns describe time and space computational complexity.

National Bureau of Standards researcher Jack Edmonds shows that algorithms of polynomial complexity are reasonably efficient, while nondeterministic polynomial (NP) problems are not.

1971 Stephen A. Cook, a researcher at the University of Toronto, develops the concept of NP-completeness.

1994 Bell Laboratories researcher Peter Shor discovers a quantum algorithm—an algorithm running on a hypothetical quantum computer—that can factor numbers in polynomial time. This means that a quantum computer could break the encryption methods commonly used on the Internet.

2000 Clay Mathematics Institute names P versus NP as one of the seven Millennium Prize Problems.

2007 Carnegie Mellon University researcher Steven Rudich and Steklov Mathematical Institute researcher Alexander A. Razborov win the Gödel Prize for showing a broad class of mathematical proofs cannot solve the question of P versus NP.

FURTHER RESOURCES
Print and Internet

Applegate, David L., Robert E. Bixby, Vasek Chvatal, and William J. Cook. *The Traveling Salesman Problem: A Computational Study.* Princeton, N.J.: Princeton University Press, 2006. This particular problem has fascinated computer scientists, mathematicians, and many other people since the early part of the 20th century. The book contains a lot of advanced material, but also includes accessible discussions on the history of the problem and how people have tried to solve it.

Appleman, Daniel. *How Computer Programming Works.* Berkeley, Calif.: Apress, 2000. Books on computer programming are plentiful. Most books explain how to program in a certain language such as BASIC or

C, but this book is more general, describing the fundamental concepts. Included are discussions of program flow, data organization and structures, operators, algorithms, and programming languages.

Cornell University. "Computational Complexity." 2005. Available online. URL: www.cs.cornell.edu/events/40years/pg38_39.pdf. Accessed June 5, 2009. This document is part of a Cornell University publication celebrating the achievements of their faculty.

Dasgupta, Sanjoy, Christos Papadimitriou, and Umesh Vazirani. *Algorithms*. New York: McGraw-Hill, 2006. This college-level text offers a comprehensive discussion of algorithms.

Dewdney, A. K. *The New Turing Omnibus*. New York: Holt, 2001. This reprint of a 1993 book presents 66 brief articles on all aspects of computer science, including algorithm analysis, artificial intelligence, data structures, logic, complexity theory, coding, and more.

Harris, Simon, and James Ross. *Beginning Algorithms*. Indianapolis: Wiley Publishing, 2005. Most books on algorithms are not written for the uninitiated beginner, and despite its title, this book is not an exception. But the authors do a good job of explaining what they mean and do not assume the reader knows too much about computer science, so a determined beginner can learn a lot about big O notation and various algorithms such as sorting and searching methods.

Lin, Shin, and Brian W. Kernighan. "An Effective Heuristic Algorithm for the Traveling-Salesman Problem." *Operations Research* 21 (1973): 498–516. The authors present a method of approximating the solution to TSPs.

Miller, Donald L., and Joseph F. Pekny. "Exact Solution of Large Asymmetric Traveling Salesman Problems." *Science* 251 (15 February 1991): 754–761. The authors describe algorithms for certain kinds of TSPs.

Spiliopoulos, Kimon. "An Introduction to Computational Complexity." Available online. URL: users.forthnet.gr/ath/kimon/CC/CCC1b.htm. Accessed June 5, 2009. This tutorial does a good job of explaining computational complexity and some of its underlying mathematics. Discussions include polynomial and exponential time algorithms, transformations, classes N and NP, and NP-complete problems.

Web Sites

Clay Mathematics Institute. Millennium Problems. Available online. URL: www.claymath.org/millennium. Accessed June 5, 2009. The scientific advisory board of the institute has deemed these seven problems to be of critical importance in computer science. One of the problems is "P versus NP," the question of whether P equals NP.

Federal Trade Commission. Identity Theft. Available online. URL: www. ftc.gov/bcp/edu/microsites/idtheft. Accessed June 5, 2009. This Web site describes the crime of identity theft and what people can do to prevent it.

Georgia Tech University. Traveling Salesman Problem. Available online. URL: www.tsp.gatech.edu. Accessed June 5, 2009. A comprehensive site that describes the history of the problem, applications, methods, and related games.

4

ENCRYPTION AND DECRYPTION: MAKING AND BREAKING CODES AND CIPHERS

Encryption, the process of scrambling a message, derives from the Greek word *kryptos,* meaning hidden, which is the same source of the word *cryptography,* referring to secret writing, and *cryptology,* the study and use of different types of secret writing. Cryptology has been at the frontier of science for a long time—a perpetual frontier, one might say, due to the constant struggle between those who wish to keep secrets and those who wish to uncover them.

Consider Sparta, for example. Sparta was an ancient Greek city-state in which the citizens prized military training. Some of the exploits of the Spartan military are legendary, such as the small group of Spartans and other Greeks who defended the pass at Thermopylae in 480 B.C.E., slowing down a huge force of invading Persians—the story, told by the ancient historian Herodotus, was the basis of the 2007 film *300.* Spartan soldiers were well-trained and loyal, and they also knew the value of maintaining secrecy. Troop movements and strategies must be kept secret, otherwise the enemy knows what to expect and will be prepared. But there must be communication among the separated elements of a fighting force, and between the commander in the field and headquarters back home, and this requires messages that might be intercepted and read by the enemy. To counter this possibility, the ancient writer Plutarch said that Spartans used a clever device known as a scytale to make their messages difficult for the enemy to read.

A scytale was cylindrical piece of wood that provided a simple form of encryption. As shown in the figure, the Spartans wrapped a thin strip of parchment around the scytale, leaving no space of the wood uncovered. According to Plutarch, Spartans wrote their message lengthwise, parallel to the cylinder's axis, then unwrapped the parchment. When unwrapped, the letters of the words were scrambled because they were no longer connected. Messengers carried this thin strip of parchment containing the message to the intended recipient. If the parchment somehow fell into the wrong hands, the enemy would not be able read it due to the scrambling of the letters. Reading the message requires that the recipient wrap the strip of parchment around another scytale of exactly the same diameter, so that the letters become connected again. Although no one knows for sure that the Spartans used scytales in this manner, the method could have been reasonably effective.

A similar desire for secrecy and privacy motivate cryptography today as much as in the fifth century B.C.E. Modern technology, such as

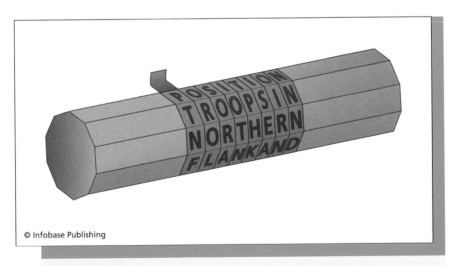

© Infobase Publishing

When unwound, the letters on the ribbon will be out of order, concealing the meaning of the message.

radio, has aided communication tremendously, but broadcasting a secret message is a bad idea unless the message is adequately encrypted. The Internet is also a concern, since the data bounces from computer to computer and can be easily intercepted. People conduct billions of dollars in transactions every year via the Internet, exposing personal information such as credit card and bank account numbers that might be stolen unless secrecy is maintained.

The advantages gained by reading someone else's secret messages have also motivated *cryptanalysis,* the process of reading or "breaking" cryptographic systems. For at least the last 2,500 years, and probably long before, a war between cryptographers and cryptanalysts has been waged. Breaking simple cryptographic systems is not difficult once the trick is known—to read a message encrypted with a scytale, for example, all the enemy needed to know was how scytales worked, and then they could find, perhaps by trial and error, a stick of the right size. Once the method has been broken, cryptographers must find a new method, which cryptanalysts will again attack and try to solve.

Now that computers are involved in most forms of technology, including communications and cryptography, cryptology is a vital and active branch of computer science. This chapter describes computer technologies used in modern cryptography, and the means by which cryptanalysts hope to break these methods some day.

INTRODUCTION

Encryption methods include the use of a *code,* in which words or phrases are replaced with symbols or other words to represent them, and a *cipher,* in which letters substitute for other letters, thereby scrambling the meaning of the words. For example, according to the ancient Roman writer Suetonius, the Roman general and politician Julius Caesar used a cipher that shifted the alphabet three letters. Applying this encryption scheme to the English alphabet, the letter *a,* the first letter, would become a *d,* which is the fourth letter, and so on for the other letters, with the final three letters—*x, y,* and *z*—wrapping around to become *a, b,* and *c* respectively. The words *seek the enemy* would be written as *vhhn wkh hqhpb.* To decipher the message, the recipient shifts the alphabet in the other direction by the same amount.

Codes use a codebook that describes what the symbols or words of the code mean. To encode a message such as "attack at dawn," the sender would look up "attack," "at," and "dawn" in the codebook, which may be represented by "robe," "@," and "zqz" respectively, so the message would be encoded as robe @ zqz. Decoding the message requires the codebook.

Many people use the term *code* in a general way, referring to both codes and ciphers. Messages without encryption are called plain text, such as "seek the enemy" or "attack at dawn," while the encrypted text is "vhhn wkh hqhpb" or "robe @ zqz." The key is the secret by which the meaning is kept hidden, and is an item of information needed to unlock the secret. In the cipher used by Caesar, the key was a shift of three letters, and in the method of the Spartans, the key was a scytale of the correct diameter.

The method used by Caesar belongs to a category of encryption known as a monoalphabetic cipher. In this cipher, each letter of the alphabet is substituted with another, and the substitution is constant.

Bar codes on packages are a code that the scanner uses to identify the item. *(Keithfrithkeith/Dreamstime.com)*

The substitution may be a shift, or some other mapping of one letter to another, but whatever the substitution method, it does not change throughout the encryption procedure. (This constancy is implied by the prefix *mono,* meaning one—in this case, a single substitution scheme.)

Recovering the plain text in a monoalphabetic cipher is easy for those who know the key. Those who do not know the key may guess at the substitution scheme by substituting letters and seeing if this turns what seems to be gibberish into an intelligible message, but this process takes time. What makes a monoalphabetic cipher secure is that the number of possible substitution schemes is huge. As described in

Combinatorial Mathematics

Mathematician Ivan Niven subtitled his 1965 book on combinatorial mathematics, *Mathematics of Choice,* with the phrase, *How to Count without Counting.* Determining the quantity can always be done by counting—increasing the total by one for each item or instance—but when the number is huge, counting takes a long time. Counting the number of substitution schemes in a monoalphabetic cipher, for example, is a prohibitively time-consuming task (and a boring one).

Combinatorial mathematics, or combinatorics, as it is often called, is a field of mathematics concerned with the arrangement or order of a set of objects. Consider the number of schemes in the monoalphabetic cipher. In the process of selecting a scheme, the cryptographer takes the first letter, say *a,* and chooses its substitution. In the English alphabet there are 26 letters, so there are 26 possible substitutions (including *a*). After this choice is made, the cryptographer moves to the second letter. Since one letter has already been selected, there are 25 remaining letters to choose from. After this selection, the next letter is chosen from 24 possibilities, and so forth, with the nth letter having $27 - n$

the sidebar, which discusses counting methods known as combinatorial mathematics or combinatorics, the number of possible substitution schemes is about 400,000,000,000,000,000,000,000,000 (4.0×10^{26}). That is a lot of possibilities! If it takes a cryptanalyst one minute to try a randomly picked possibility, he or she would have to get lucky to find the right one even after millions of years.

Because there are so many possibilities, monoalphabetic ciphers would have been unbreakable in the age before computers—except for one thing, first noted by Arab scholars such as Yaqub Ibn Ishaq al-Kindi in the ninth century. These early cryptanalysts noticed that certain let-

possibilities. Even if the cryptographer adopts a scheme such as a shift or some other simple operation, the concept and the result are the same—selections from a number of possibilities.

A basic principle of combinatorics is that the total number of something equals the product of the number of possibilities. For example, if a person has a choice of two different shirts—red (r) and blue (b)—and three different pairs of pants—gray (g), brown (w), and blue (b)—there are $2 \times 3 = 6$ total wardrobe combinations—r-g, r-w, r-b, b-g, b-w, and b-b. (Some of these combinations would fail to be fashionable!)

For a 26-letter cipher, the cryptographer has 26 possibilities for the first choice, 25 for the second, 24 for third, and so on. The product of these numbers is $26 \times 25 \times 24 \times 23 \times 3 \times 2 \times 1$. The product of all positive integers below n is called n factorial, represented mathematically by an exclamation mark—$n!$. Factorials are huge for even small n. For example, 26! is approximately 4.0×10^{26}. It includes all schemes, including the undesirable one (from a cryptographer's point of view) of choosing a for a, b for b, and so on for each letter of the alphabet. More advanced methods of combinatorics deal with problems of selection, but even without these poor choices, the number of ways of picking a monoalphabetic cipher is enormous.

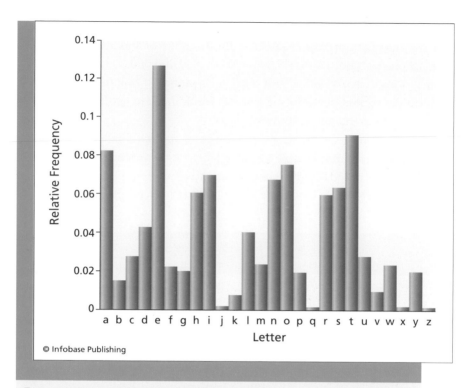

© Infobase Publishing

For each letter of the English alphabet, plotted on the horizontal axis, the relative frequency is shown on the vertical axis. For instance, the letter *a* appears in words with a frequency of about 0.08. This means an average of 8 percent of the letters in an English message are an *a*.

ters appeared more frequently than others, and other letters appeared most often in certain combinations. For example, *e* is the most commonly used letter in the English alphabet, and the letter *q* is rarely followed by any other letter than *u*. If the cipher for "seek the enemy" is "vhhn wkh hqhpb" as given above, a clever cryptanalyst might guess that the letter *h* in the encrypted text stands for *e* since it appears frequently. This kind of analysis is known as *frequency analysis*. The figure shows an example of a frequency analysis chart.

Crossword puzzle lovers know that once a few words are filled in, the puzzle gets easier. The same is true for deciphering encrypted text—once a few letters are known, most of the rest of the text can be guessed. Frequency analysis works for monoalphabetic ciphers but does not work on well-designed codes, since the symbol or word assignments often have little if any relationships. But codes are more complicated than ciphers and suffer from a serious defect, which will be discussed in more detail later in the chapter—the need for a codebook to decode encrypted text. Codebooks must be distributed to all of a message's recipients or they will be unable to read the message, and this widespread distribution offers the enemy plenty of opportunities to steal one. In most cases, ciphers have been the preferred means of encryption. With the Arabic invention of frequency analysis, cryptanalysts had a tool to crack most monoalphabetic ciphers, giving them the upper hand in the arms race, so to speak, between cryptographers and cryptanalysts.

But a new and improved cipher scheme emerged in the 15th and 16th centuries, during the Renaissance period. The idea was to use a number of different alphabet ciphers for the same message—a polyalphabetic cipher. Imagine 26 ciphers, with shifts of 1 through 26; a shift of 1 means *b* stands for *a, c* for *b,* and so on, and a shift of 2 means *c* stands for *a, d* for *b,* and so on, all the way down to a shift of 26, in which the alphabet is unchanged. The set of these 26 shifts form a type of polyalphabetic cipher known as a Vigenère cipher, named after Frenchman Blaise de Vigenère (1523–96), who published it in 1586.

To encipher the message, choose a word, such as *keyword,* and the message to be enciphered, such as "Meet me in Memphis." The key, in this case *keyword,* determines which cipher is to be used for each letter of the message. For the first letter of the message, use the K shift, because K is the first letter of the key—the K shift is the one where *k* stands for *a, l* stands for *b,* and so on. In the K shift, *w* stands for *m,* so the first encrypted letter of the message is *w.* The second letter of the message uses the E shift because E is the second letter of the keyword, and in this cipher *e* stands for *a, f* stands for *b,* and so on. The second letter of the plain text is *e,* so the second letter of the encrypted message is *i.* The third letter uses the Y shift, a cipher in which *c* stands for *e.* The encrypted message thus far is "wic." This process continues until the message is completely encrypted. If the message is longer than the key, as is the case in this example, the cryptographer reuses the letters of the key, in the same order as before. The encrypted message will be "wicpavlxqcidylc" (omitting the spaces). To decipher the text, the recipient works backward—but he or she needs the key in order to know which cipher to use for each letter.

Notice that a cryptanalyst attempting to use frequency analysis has a problem. Letters are encrypted with several different ciphers, so that the encrypted word for "meet" is "wicp"—the two instances of *e* do not use the same letter. The number of possibilities is even larger than with a monoalphabetic cipher, and cryptanalysts are deprived of easy guesses.

Polyalphabetic ciphers seemed unbreakable. But once again, this assumption proved wrong. British engineer Charles Babbage (1791–1871), who played a role in the development of computers, and Friedrich Kasiski (1805–81), a retired Prussian army officer, independently cracked the cipher in the middle of the 19th century. To do so, they analyzed the complicated structure created by reusing a key, such as *keyword,* repeatedly for a long message. Because language has certain patterns, such as those used in frequency analysis, there will be patterns in the encrypted text, although in polyalphabetic ciphers this structure is usually not apparent without considerable effort on the part of an expert cryptanalyst.

Since the patterns occur because of structure in the key and its repetitive use to encrypt a lengthy message, what if the operator used a key as long as the message? There would still be structure in the key, except when the key consists of random letters. Random keys that are as long as the message (so that the key is not repeated) removes all structure from the encrypted text. Ciphers that use random keys with no structure are sometimes called onetime pad ciphers, introduced by U.S. army officer Joseph Mauborgne (1881–1971) and engineer Gilbert Vernam (1890–1960) in 1917. Onetime pad ciphers are unbreakable, and this is not just wishful thinking by cryptographers—there is no structure or patterns in the encrypted text if the procedure is correctly followed, which makes this cipher truly unbreakable without the key.

The problem with using long keys of random letters is that they cannot be memorized. If an army sent its units into the field, each unit needs to know the key in order to decipher encrypted messages. If the key is short and memorable, such as *keyword* or *yankee doodle dandy,* operators would have no trouble remembering it.

But long keys consisting of random letters must be written down or they will not be remembered. The written record creates the same problem that plagues codes—with plenty of copies of the key lying around, sooner or later the enemy is bound to discover or capture one. The key must be changed at once if this happens or the encryption is useless. And if the enemy secretly steals a key that continues to be used, no secrets are safe. A code is effective only if the codebook is secret, and a cipher is effective only if the key is secret. In his 1999 book *The Code Book,* Simon Singh wrote, "A onetime pad is practical only for people who need ultrasecure communication, and who can afford to meet the enormous costs of manufacturing and securely distributing the keys. For example, the hotline between the presidents of Russia and America is secured via a onetime pad cipher."

People used Vigenère ciphers to encrypt telegraph messages and other correspondence in the 19th century even though the cipher could be broken—it was good enough to baffle anyone except a practiced cryptanalyst. But the stronger form of encryption was necessary to maintain absolute secrecy.

The development of radio in the 1890s by Italian physicist Guglielmo Marconi (1874–1937) and others initiated a new age of communication—and a new age of encryption. Radio makes it easy to contact people, but unintended as well as intended recipients can hear the message. To combat this loss of privacy, people had to develop new and improved encryption devices.

ENIGMA—AN ENCRYPTING MACHINE

Onetime pad ciphers are unbreakable because randomness eliminates any detectable pattern in the enciphered text. If a machine could be invented to produce a cipher so scrambled as to appear random, then it would generate little if any structure that cryptanalysts could exploit. In 1918, German engineer Arthur Scherbius (1878–1929) invented such a machine, known as Enigma. The German military later adopted this machine to encrypt many of their radio communications during World War II (1939–45).

Enigma machines used rotary dials and complicated circuits to encipher the letters of the message. As the operator typed each letter, the machine's complex internal circuitry chose the enciphered letter, then automatically advanced the dials and switches to a new cipher scheme for the next letter. This is a polyalphabetic cipher, in which the machine's circuitry selects the ciphers instead of using a keyword. Since a group of machines were made with exactly the same circuitry, they all enciphered a message the same way provided they began in the same initial position, which was determined by certain settings. The machines could also work in reverse, so that they would be able to decipher a message that any of the machines in the group had enciphered, as long as the machines had the same settings.

Enigma machine *(The National Security Agency/Central Security Service)*

The problem is the settings. Suppose a fleet of submarines is equipped with the same type of Enigma, and the settings are initially the same on each machine. The submarines could in principle use the machines to encipher and decipher messages, but the settings must not be changed on any of the machines, otherwise they would become unsynchronized. This means the machines are continually going through the same progression of ciphers, and if there a lot of messages to be sent—which can amount to millions of words every week—a clever mechanical engineer might be able to figure out how the circuitry works by studying the encrypted messages. To avoid this disaster, operators need to change the settings daily. Since the machines must stay synchronized, every operator must make the same change. These settings become the keys of this encryption scheme, and like all keys, they must remain secret. But every operator has to have a copy of the settings each day—and the problem of keeping the keys safe from prying eyes appears yet again.

The distribution of keys was a vulnerability benefiting the Allies—Great Britain, France, the United States, and Russia—during World War II. Stealing these keys allowed listeners to decipher Engima messages. But Allied cryptanalysts could not always rely upon the ability of secret agents to steal keys. Instead, cipher-breakers took advantage of guesses, along with a lot of mathematical ingenuity, to crack Enigma.

For instance, a German weather station may encrypt its daily report with Enigma. Allies listening to the transmission would pass the encrypted message to cryptanalysts, who could guess at some of the words the report would contain, such as temperature and humidity. These guesses often provided important clues. In addition, mathematicians studied Enigma messages to discern highly complicated patterns present in the enciphered text—the machine did not function in a completely random fashion. British cryptanalysts included Alan Turing, who was also the developer of the Turing test, and who made other important contributions to computer science in the 1930s and 1940s. Working under top-secret conditions from Bletchley Park—an estate in Bletchley, England—Turing and his colleagues improved upon techniques first developed to crack Enigma a few years earlier by Polish mathematician Marian Rejewski (1905–80) and his colleagues at Poland's cipher bureau, Biuro Szyfrów. Turing directed the construction of gigantic calculating machines known as bombes, which began arriving at Bletchley Park in 1940.

Thanks in part to stolen keys and in part to the mathematical deciphering of Turing, his colleagues, and the bombes, the Allies could read many of the Germans' secret messages. Confident that Engima was secure, Germany kept using it, resulting in serious information leaks. No one knows exactly how much the Enigma cryptanalysts aided the Allied cause, but British Prime Minister Winston Churchill (1874–1965) made this effort a top priority. Having access to the opponent's plans was a huge advantage, especially during D-day, the successful Allied invasion of German-occupied France on June 6, 1944.

COMPUTERS AND CRYPTOLOGY

Enigma was a remarkable machine. Its cipher would have been secure except for stolen keys and a great deal of effort by some of the world's brightest people, such as mathematician and computer scientist Alan Turing. The exploitable weaknesses of the Enigma machine were partly due to the problem of key distribution, and partly due to the limits of the electrical and mechanical components. Despite its advanced circuitry, Enigma failed to remove all the structure in its encrypted messages.

Computers are different. A computer is a general-purpose machine that can run a variety of programs, including programs to produce ciphers and codes of unfathomable complexity. The same concepts described earlier in the chapter apply to cryptography with computers, but computers are so fast that the algorithms can be much more complicated. The rise in computer use in the 1960s initiated the age of computational cryptography. Computers, when properly programmed, give users an excellent tool to encrypt messages and maintain the privacy of their communications, and computer programmers have developed many encryption algorithms.

Recall that computers function with a language of binary numbers—1s and 0s. A bit is one unit of information, either a 1 or 0, and computer data and instructions consist of strings of bits representing numbers or commands.

When corporations and governments began computerizing their daily operations in the 1960s, there were many types of computers and software, and few standards or common procedures existed among the different varieties. Sending encrypted messages within a company or branch of government was not a problem because everyone was using the same computers and algorithms, but sending encrypted messages to another organization presented a difficulty. Since each business or government office had its own encryption techniques and algorithms, communication between them was not possible. The recipient must have the key and how it is used—the algorithm—in order to decipher the message.

Resolving the ensuing chaos required the adoption of a standard—a common technique or measurement that everyone knows how to use. Adopting a standard is similar to using the same language, which allows all speakers to communicate with one another. In 1976, the United States government approved the Data Encryption Standard (DES), employing an algorithm developed by mathematicians and scientists at IBM.

DES encrypts blocks of text based on a key. The algorithm was not secret—people had to know how to implement it on their computers so that it could be used by anyone who wanted to. The secrecy of the encrypted text depended on the secrecy of the key. This is the same as earlier encryption methods, where the cipher technique is known to all, and privacy relies on limiting the knowledge of the key to the sender and recipient. The only new feature at this point was the complicated nature of the algorithm, which was designed to prevent patterns such as those that spoiled Enigma's encryption scheme.

When using computers for cryptography, the keys are numbers in binary form. There should be many possibilities to prevent a cryptanalyst from guessing the key. In its original implementation, DES had a 56-bit key, which means that the key was composed of 56 bits.

How big is a 56-bit key? Each digit or bit of a binary number represents a power of two, just as each digit of a base-10 number represents a power of 10. In the commonly used base-10 system, the first digit represents ones (10^0), the second tens (10^1), the third hundreds (10^2), and so on. For the binary system, the first digit represents $2^0 = 1$, the second represents $2^1 = 2$, the third represents $2^2 = 4$, and so on. Thus 101 in binary equals 5 in base-10, and 011 equals 3. A 56-bit number can equal a base-10 number as low as 0, when all bits are 0, a number as high as 72 quadrillion (72×10^{15}) when all bits are 1. A 56-bit key means that there are 72 quadrillion possibilities.

Computers are fast, so encryption algorithms can be complex, but computers can also be programmed to decrypt by trying guesses, speeding up the cryptanalyst's work. People began to wonder if 56 bits was enough. This number of keys was lower than the possible ciphers of the old monoalphabetic cipher, which had been broken long ago. But an important factor was the lack of structure or patterns in a correctly performed encryption algorithm, which did not allow frequency analysis or other guessing schemes to work. Breaking a perfect encryption algorithm would require brute force, checking each possibility one at a time. Even computers, at least in the 1970s, were not fast enough to succeed in a reasonable amount of time.

There were other issues as well. Making a powerful encryption scheme available to anyone also makes it available to thieves, terrorists, and spies. As described in the following sidebar, the United States cryptology agency, the National Security Agency (NSA), is responsible for protecting government information as well as collecting intelligence from the transmissions and communications of potentially hostile organizations. The agency's mission involves a considerable amount of cryptography and cryptanalysis. Government officials and the IBM scientists who developed DES deemed a 56-bit key to be sufficient protection for normal circumstances, since in 1976 most computers could not reasonably be expected to succeed in a brute-force analysis. But the nearly inexhaustible resources of the United States government might have succeeded if necessary, which could have given NSA a chance to decipher messages. This would not have been the case if the standard had employed a larger key.

NSA may or may not have been able to read DES with 56-bit key when it was first adopted—few people outside of the agency can be sure, since NSA is not in the habit of divulging secrets—but advances in computer

National Security Agency

The success of Great Britain's Bletchley Park in cracking Enigma during World War II was duplicated in the Pacific theater of the war. United States military intelligence units broke several important Japanese codes and ciphers, including one used by the Japanese navy. Knowledge of Japanese strategy was instrumental in several American victories, including the defeat of the Japanese navy in the Battle of Midway in June 1942, a decisive turning point in the war. During World War II, military cryptanalysts worked for separate units or defense agencies, but in 1949 these efforts coalesced into one agency, the Armed Forces Security Agency. Officials quickly recognized the value of coordinating the efforts of all cryptologists, including military and nonmilitary operations, so on November 4, 1952, an overall security agency, the National Security Agency (NSA), was established.

NSA headquarters is located at Fort Meade, Maryland, which is between Baltimore and the nation's capital, Washington, D.C. The agency has about 30,000 employees worldwide, with about half military and half civilian. A large variety of specialists make up this substantial workforce, including computer scientists, mathematicians, physicists, engineers, and linguists. The need for highly qualified cryptographers and cryptanalysts means that NSA is a particularly large employer for skilled computer scientists and mathematicians. The agency also conducts cryptology research, and sometimes provides funds to cryptologists outside the agency to conduct special projects.

NSA is one of 15 federal organizations comprising the United States Intelligence Community—the departments and agen-

technology and speed made a 56-bit key obsolete anyway. A standard known as Advanced Encryption Standard (AES) superseded DES in 2001. The key size in AES can be up to 256 bits, a number exceeding 10^{77}.

National Security Agency building *(National Security Agency)*

cies tasked to gather and protect sensitive information. The list includes the Central Intelligence Agency and the Federal Bureau of Investigation. These agencies are known for their secrecy—a nickname for NSA is "Never Say Anything"—which sometimes leads to suspicions or allegations of spying on American citizens, even though the Fourth Amendment of the United States Constitution bans "unreasonable searches and seizures." Legality of the activities of these agencies and what constitutes "unreasonable searches" are issues to be decided in the courts, but a cryptology agency is essential to national defense, as clearly indicated by the events of World War II.

But there is still a vulnerability—the keys must be distributed to recipients, but keys must be kept hidden from everyone else. The complicated algorithms remove most of the patterns from the encrypted text, yet the cipher can be cracked easily if someone steals the key.

The problem of key distribution led some computer scientists to consider the possibility of encryption algorithms that do not rely on the exchange of secret keys. This research led to a new kind of cryptography that is widely used today.

PUBLIC KEYS

An encryption procedure in which the sender and recipient use the same key, such as the procedures described above, is called symmetric cryptography—the key is the same at both ends of the communication channel, which gives the procedure its symmetry, since the operation of extracting the message is the same, except in reverse, as that used to scramble it. In 1976, Stanford University researchers Martin Hellman, Whitfield Diffie, and Ralph Merkle demonstrated a type of asymmetric cryptography, in which the operation to encrypt the message is not the same as the one to decrypt it. The encryption key does not have to be kept secret so it can be openly exchanged, in public, between sender and recipient without worrying about its security. Public keys avoid the problem of key distribution by eliminating the need to keep the keys secret.

Public key or asymmetric cryptography actually uses a pair of keys, one of which must be exchanged—this is the public key—and one of which is private. The private key is only needed for *decryption*—reading the message—not encryption, so there is no need for the sender to know it. For example, suppose someone wishes to send a secret message to X. To encrypt the message, the sender uses a public key, a key that X has published. When the message is received, X uses the private key—which is carefully guarded—to decrypt the message. If X wishes to reply to the sender with an encrypted message, he or she uses the sender's public key. Anyone can encrypt a message to anyone else as long as they both know the public key, but decryption relies on the private key.

Asymmetric cryptography works because the public and private keys are mathematically related. The encryption algorithm is common to all users, and the sender (or the sender's computer) uses a number, the public key, to encrypt the message. Undoing this encryption requires another number, the private key, of which there are many possibilities, so that no one will be able to guess the private key and decipher the message.

In order to work, the asymmetric, two-part process needs two numbers that are mathematically related, so that one is a function of the other; it can be computed in terms of the other, such as $y = e^x$, where x and y are numbers and e represents the exponential function. The public key is known to everyone, but the private key must remain confidential, which means that the relationship cannot be easily guessed.

Asymmetric cryptography relies on one-way functions, which are easy to compute in one direction but not in the other. The example above, $y = e^x$, is not an effective one-way function because it is easy to find y (simply raise e to the power of x), and easy to find x from y (take the natural logarithm of both sides of the equation to find x—the natural logarithm of e^x is x, so x equals the natural logarithm of y). But exponential functions in modular arithmetic, in which a group of numbers forms a loop,

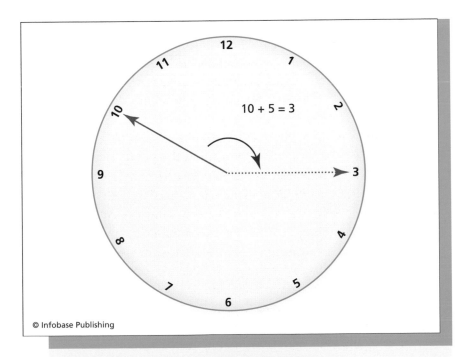

In modular arithmetic with modulus 12, 10 + 5 = 3.

can be effective one-way functions. Modular arithmetic is also sometimes called clock arithmetic, since it is related to 12-hour clock time in which the modulus is 12. For example, adding five hours to 10 o'clock makes three o'clock, because in mod 12, 10 + 5 = 3, as illustrated in the figure. Factorization, described in chapter 3, is also important.

Soon after Hellman, Diffie, and Merkle came up with their public key concepts, a team of researchers at the Massachusetts Institute of Technology—Ronald Rivest, Adi Shamir, and Leonard Adleman—explored these functions and published their findings in 1978. The asymmetric cryptography method based on their work became known as RSA, for the initials of the last names of the three researchers.

Public key cryptography allows anyone to send anyone else a message, as long as the sender knows the recipient's public key. RSA users publish their keys for this very purpose. Messages are encrypted with the public key, and the recipient decrypts these messages with the private key. The public key comes from a one-way function, so it is almost impossible to use it to discover the private key and decrypt the message. Only the recipient can decrypt the text. The vulnerability of distributing and exchanging secret keys is therefore avoided.

Reversibility of a one-way function must be computationally difficult in order to avoid the discovery of the private key by speedy computers. Many problems require a prohibitively long time to solve as the amount of data gets large. In the case of asymmetric cryptography, the relationship between private and public keys involves factorization of integers, which is not quite one of the hardest problems known, but still complex enough to be beyond today's machines to compute when the numbers are extremely large.

Asymmetric cryptography is one of the most important tools in modern cryptology. Singh wrote in *The Code Book,* "The great advantage of RSA public key cryptography is that it does away with all the problems associated with traditional ciphers and key exchange." RSA and similar algorithms permit anyone to send encrypted messages without first having to exchange a key, which can be intercepted and stolen. This is especially important in an environment such as an extensive network with many users, in which keys would have to be widely distributed.

Yet asymmetric cryptography is not without its faults. It is generally slower than symmetric cryptography because the computer must perform a lot of mathematical operations. Sometimes symmetric cryptography remains the best solution, particularly when the number of users is limited and the distribution of the secret keys is not quite so dangerous.

Believing that public keys would benefit users of electronic mail (e-mail), political activist and computer scientist Philip Zimmermann devised a speedier version of the algorithm in 1991 and incorporated it into an e-mail encryption program. The program, called Pretty Good Privacy (PGP), succeeded so well that Zimmermann became the target of a federal investigation. The United States government launched the investigation because PGP had become available globally, but exporting American technology capable of producing difficult-to-break encryption was restricted in the 1990s due to fears that NSA would be unable to monitor terrorists' communications. PGP used 128-bit keys.

Exportation restrictions have since been relaxed, and Zimmermann was never charged with a crime. Today PGP remains one of the most popular e-mail encryption programs. (Zimmermann derived the name of his computer program from the name of a market, Ralph's Pretty Good Grocery, featured in comedian Garrison Keillor's stories and broadcasts.)

The asymmetry of public key cryptography is also useful when the procedure is reversed. A person can use his or her private key to encrypt a message, which recipients can decrypt with the public key. This process clearly offers no security, since everyone has access to the public key and can read the message. Instead of sending secret messages, encrypting with a private key is a method of ensuring that a message came from a certain person, namely, the person who owns the private key. Because private and public keys are mathematically related, a message that can be decrypted with a person's public key means that it was encrypted with his or her private key. This technique provides a "digital signature," an electronic version of a signature that identifies the message's author.

Although asymmetric cryptography solves the key distribution problem, it introduces another problem. Suppose someone publishes a public key claiming to be from a certain person, but the public key is actually not from that person at all. Encrypted messages using this fraudulent public key will be readable by the person who has the associated private key. To prevent this, asymmetric cryptography users often have their public keys verified by a trusted company or third party, called certification authorities.

Identity theft and message interception are two major concerns with Internet usage. Billions of dollars of business transactions pass along the network of connected computers that make up the Internet, potentially exposing a great deal of private information. One of the most important applications of the field of cryptography is to ensure the security of this information.

CRYPTOLOGY AND THE INTERNET

Internet transactions such as the purchase of books or applications for loans usually require the customer or applicant to submit important personal information, such as a credit card number or a Social Security number. Identity theft is a crime that occurs when someone fraudulently poses as another person in order to use that person's credit, assets, or benefits. Common identity theft crimes include the use of stolen credit card numbers to make purchases, and the use of stolen Social Security numbers to apply for loans or benefits. The network of computers making up the Internet is not secure or private—information passes from computer to computer as it travels to its destination, and it can be intercepted easily in route. To keep information private, Internet transactions must be encrypted.

Transactions conducted over the Internet "leak" information, which must be encrypted in order to protect the user's privacy. *(Victor Habbick Visions/Photo Researchers, Inc.)*

Most of the time, computer programs automatically perform the necessary encryption, employing the methods discussed above. But Internet browsers such as the Microsoft Corporation's Internet Explorer will not normally encrypt information unless the user visits a Web site with a secure link. These sites are identified with a Web address beginning with *https* rather than *http.* A procedure known as a protocol governs the manner by which the browser, running on the visitor's computer, interacts with the Web server, which is the computer hosting the Web page. An older version of this protocol is known as Secure Sockets Layer (SSL), and a newer version is called Transport Layer Security (TSL). The protocol specifies that the computers follow a specific encryption algorithm such as RSA, the selection of which depends on the algorithms the computers are capable of running. The cryptography is asymmetric, so the computers create a public and private key with which they will encrypt and decrypt messages. These functions are "transparent" to the computer user, meaning the operations are handled by the computer.

Many browsers indicate SSL or TSL connections with an icon positioned somewhere on the page. Internet Explorer uses a gold padlock, which in the 7.0 version appears near the top, to the right of the address (on previous versions of this browser the padlock appeared near the bottom). The padlock is a visual indication that the connection between the client computer (the computer on which the browser is running) and the Web server is secure. In some cases, only a portion of a page is secure; for instance, a Web page may have a log-in program that establishes a secure connection, in which case the gold padlock does not appear on the page.

But any cryptography system is only secure if the user is careful. Criminals can write computer programs that install themselves on a person's computer without the owner's consent. For example, viruses and worms are malicious programs that travel along computer networks in e-mails or other network traffic, infecting computers that are not protected with antivirus software. These programs may have instructions for searching a person's computer files for personal information, including private keys. Computer users who are connected to the Internet should exercise caution when opening e-mails from unknown persons or when visiting unfamiliar Web sites.

Another common technique known as *phishing* has emerged recently to bypass cryptographic protection. The technique, whose name resembles the word *fishing,* is an attempt to get unwary computer users to visit a Web site and provide personal information. The Web site may appear to be from a legitimate bank, company, or other institution, and the connection may also be secure, but the Web site is actually phony, designed specifically for the purpose of collecting credit card numbers or other private data. Criminals entice users to visit these sites by providing links in fraudulent e-mails that for instance, may claim to be from a bank or company requesting personal information for some reason, such as updating its files. When users click on the link, they are taken to the phony Web site.

To avoid phishing scams, computer users must be careful about using links to visit important Web sites. If a link appears doubtful, a user can type in the correct address, rather than trust the link. At secure Web sites, identified with an icon such as the gold padlock on Internet Explorer, users can usually double-click on the icon to reveal Web site information provided by certification authorities, who verify Web sites as well as cryptography keys. The information should match the Web site's owner and address.

E-mail is not generally encrypted unless the user installs a program such as PGP. The Internet community may eventually opt for standard encryption of e-mails, however, as more and more people use e-mail for important correspondence.

Tactics such as viruses and phishing are not attacks on the encryption algorithm itself, but rather attempts to trick a person into revealing secrets. These attempts are similar to stealing codebooks or keys, as secret agents did during World War II.

Encryption tools such as the Advanced Encryption Standard and public key cryptography are hard to break, but the rapid increases in computer processing speed mean the chances for success of a determined cryptanalyst are getting better. Much of the research on the security of today's encryption algorithms involves the field of computational complexity, the subject of chapter 3. This field of computer science studies the efficiency of algorithms and the problems they are capable of solving in a reasonable period of time. Encryption algorithms are designed to be extremely tough to solve, even with the fastest computers.

But no one has found any proof that efficient solutions do not exist for even the most difficult problems, even though researchers have spent years searching for and failing to find such solutions. And certain methods, such as those known as differential cryptanalysis, probe for patterns, where the computer encryption algorithm fails to be random. Such patterns are similar to those found in the Enigma machine's encrypted messages by Rejewski, and later exploited by Turing and his colleagues during World War II.

The most dangerous threat at present to the security of modern cryptographic systems is probably not a cryptanalysis technique, but a new kind of computer on the horizon. Quantum computers, described in chapter 1, would employ quantum mechanics—the physics of small particles such as atoms—to make amazingly fast calculations. Although quantum computers do not yet exist, the early phases of research into these machines have been successful enough to make cryptographers take notice.

QUANTUM CRYPTANALYSIS

In 1994, Peter Shor, a researcher at Bell Laboratories in New Jersey, found an algorithm for quantum computers that could quickly solve the difficult mathematical problem of reversing the one-way functions used in asymmetric cryptography such as RSA. Another Bell Laboratories researcher, Lov Grover, discovered a different algorithm for quantum computers in 1996 that could threaten any and all encryption schemes. This algorithm performs fast searches.

Searching is a common task for computer programs. Examples include finding a certain name on a list, or finding the owner of a car with a certain license plate in a list of registered vehicles. A short list, say 12 items, requires little time to scan, but a long list, say 1,000,000 items, can be very time-consuming. The searcher might get lucky and find the right item on the first page of the list, or the searcher might get unlucky and find the right item on the last page, after plowing through 999,999 wrong items. On average, about 500,000 items must be examined to find the right one in a 1,000,000-item list.

Algorithms for these search tasks are not nearly as inefficient as the algorithms for NP problems, but a lot of time is required for especially long lists. This great amount of time is what cryptographers count on when they use encryption schemes with a quadrillion possible keys. Cryptanalysts would have to try an average of about one-half of these possibilities before finding the right one, which would take far longer than the lifetime of the cryptanalyst, even using today's fastest computer. In general, for N items, an average of N/2 steps must be performed using conventional algorithms.

But Grover's algorithm, running on a hypothetical quantum computer, could perform the same search in much less time. Instead of N/2 steps, the quantum algorithm needs an average of only about $\sqrt{N}$ (the square root of N) steps. For 1,000,000 items, only $\sqrt{1,000,000} = 1,000$ steps would be required, not 500,000.

Grover's quantum algorithm frightens cryptographers because it would make a brute-force attack—a search for the right key—feasible against any encryption technique. With only the square root of the number of possible keys to examine, even huge lists can be searched efficiently.

Cracking ciphers and codes with quantum computers is still in the future, and cryptographers are unsure how soon it will arrive, if ever. But John Chiaverini, a researcher at the National Institute of Standards and Technology, and his colleagues recently demonstrated a small-scale version of a quantum computer that could perform one of the crucial steps in Shor's algorithm. This step, which finds patterns, is related to a mathematical operation known as a Fourier transform, named after French mathematician Jean Baptiste Joseph Fourier (1768–1830). Conventional computers, which use "classical" physics instead of quantum mechanics, require a long time to perform these operations because they involve a large number of steps. Chiaverini used a little bit of classical physics along with three quantum bits (qubits) made of beryllium ions. The researchers published their report, "Implementation of the Semiclassical Quantum Fourier Transform in a Scalable System," in a 2005 issue of *Science*. The development of an accurate quantum mechanical version of the Fourier transform is "a necessary step toward large number-factorization applications of quantum computing," as the authors noted. Such speedy factorizations would threaten the security of many encryption schemes.

The consequences of a fully operational quantum computer, should one be perfected, are enormous. Encryption that protects Internet commerce, personal information, military communications, and sensitive government projects would be vulnerable. Further research on quantum cryptanalysis and quantum technology may force cryptology to change its methods once again.

CONCLUSION

The history of cryptology is one of a fierce battle between cryptographers who develop codes and ciphers, and cryptanalysts who try to break them. Early monoalphabetic ciphers, such as Caesar's shift, seems simple to the modern cryptologist, but in their day these ciphers baffled the best cryptanalysts. After the rise of frequency analysis, which broke simple monoalphabetic ciphers, and the pattern analysis methods of Babbage and Kasiski that broke the more complicated polyalphabetic ciphers, cryptanalysts had the advantage. But the tide turned when computers came along in the middle of the 20th century, and sophisticated algorithms and encryption standards have rendered cryptanalysis much more difficult. And public key cryptography has reduced the opportunities of thieves who steal keys.

The battle continues. Quantum computers and the algorithms developed by Shor and Grover are promising tools for cryptanalysts. The potential of quantum computation is causing concerns about the future security of the encryption algorithms that safeguard a great deal of sensitive information on the Internet and elsewhere.

But quantum computation is also a potential tool for cryptographers as well as cryptanalysts. Onetime pad ciphers are unbreakable because they are completely random. The problem with this encryption method is the distribution of the lengthy keys, which must be written down, and can therefore be copied or stolen. Quantum key distribution, first proposed in 1984 by IBM researcher Charles H. Bennett and University of Montreal researcher Gilles Brassard, is a technique employing quantum mechanics to make the key exchange between sender and recipient as secure as possible. Although the technique is in its early stages of development, officials in Switzerland used it in October 2007 to secure the communication lines involved in counting votes in national elections.

Quantum key distribution is secure because one of the important principles of quantum mechanics allows users to detect attempts to steal the keys. Until someone makes a measurement, the state of a particle is a superposition of possible states, and the state of a system of particles in quantum mechanics exists in a fuzzy, intertwined combination called entanglement. The process of making a measurement determines the state. Quantum key distribution uses superposition and entanglement to transmit random keys, and any attempt to steal the key must involve measurement, which detectably alters the state. If Ed and Ted exchange quantum keys but Bob intercepts and steals the key, Ed and Ted will detect the compromise and change keys.

Randomness and foolproof key exchange would seem to be unbeatable. Yet other codes and ciphers that were once considered unbreakable have been broken, much to the shock and grief of those using them. Secrecy has not lost any of its importance in this age of computers, and the war between people who develop codes and ciphers and the people who try to break them rages on. Cryptology is a continuing frontier of computer science protecting personal information, electronic transactions, Internet commerce, military secrets, and perhaps in the future, even the integrity of elections.

CHRONOLOGY

ca. 500 B.C.E.	Spartans send encrypted messages with scytales, according to the ancient writer Plutarch.
ca. 50 B.C.E.	Roman general and politician Julius Caesar uses a monoalphabetic cipher to encrypt secret messages.
ca. 900 C.E.	Arab scholars invent frequency analysis to break monoalphabetic ciphers.
1467	Italian scholar Leon Battista Alberti (1404–72) invents the first known encryption device, consisting of concentric dials for encrypting and decrypting alphabetic ciphers. Alberti also develops a polyalphabetic cipher.
1553	Italian scholar and adventurer Giovan Battista Bellaso (1505–?) publishes a version of the polyalphabetic cipher similar to what is known as Vigenère's cipher.
1586	French cryptographer Blaise de Vigenère (1523–96) publishes a stronger form of the polyalphabetic cipher that now bears his name.
1850s	British engineer and computer pioneer Charles Babbage (1791–1871) discovers a method of break-

ing Vigenère's cipher, though he did not publish his work.

1860s   Friedrich Kasiski (1805–81), a retired Prussian army officer, independently discovers a method of breaking Vigenère's cipher, and publishes his work in 1863.

1917   American army officer Joseph Mauborgne (1881–1971) and engineer Gilbert Vernam (1890–1960) develop the unbreakable onetime pad cipher.

1918   German engineer Arthur Scherbius (1878–1929) invents Enigma, which the German military will eventually use to encrypt its communications in World War II.

1930s   Polish mathematician Marian Rejewski (1905–80) and his colleagues at Poland's cipher bureau, Biuro Szyfrów, discover methods of partially breaking Enigma encryption.

1940s   British mathematician and computer scientist Alan Turing (1912–54) and his colleagues, working at Bletchley Park in England, enhance Rejewski's techniques. Along with an occasional stolen key, the cryptanalysts succeed in breaking Enigma encryption.

1952   The National Security Agency (NSA), an agency of the United States government involved in protecting and collecting sensitive information, is established.

1976   The United States government adopts the Data Encryption Standard (DES).

         Stanford University researchers Martin Hellman, Whitfield Diffie, and Ralph Merkle demonstrate public key (asymmetric) cryptography.

1978   Massachusetts Institute of Technology researchers Ronald Rivest, Adi Shamir, and Leonard Adleman publish an implementation of public key cryptography known as the RSA algorithm.

1991   Philip Zimmermann releases Pretty Good Privacy (PGP), an encryption tool with public key cryptography that is fast enough to be used by individuals on personal computers.

1994   Bell Laboratories researcher Peter Shor discovers an algorithm for quantum computers that could efficiently reverse the one-way functions used in asymmetric cryptography such as RSA, thereby breaking the encryption—if and when quantum computers are developed.

Netscape Communications Corporation develops the Secure Sockets Layer (SSL) protocol.

1996   Bell Laboratories researcher Lov Grover discovers an algorithm for quantum computers that drastically reduces search times, which opens the possibility of successful cryptanalysis of almost any encryption algorithm if and when quantum computers are developed.

2001   The United States government adopts the Advanced Encryption Standard (AES).

2005   John Chiaverini and his colleagues at the National Institute of Standards and Technology demonstrate a small-scale quantum computation to implement one of the steps in Peter Shor's algorithm.

2007   Officials in Switzerland use quantum key distribution to secure computer networks participating in counting votes during national elections.

FURTHER RESOURCES
Print and Internet

Budiansky, Stephen. *Battle of Wits: The Complete Story of Codebreaking in World War II.* New York: Touchstone, 2002. Cryptography played a vital role in World War II. This book tells the stories of Turing and his colleagues as they broke the Enigma cipher, the American cryptanalysts who broke the Japanese navy's ciphers, and other cryptanalysis efforts.

Chiaverini, J., J. Britton, D. Leibfried, E. Knill, M. D. Barrett, R. B. Blakestad, et al. "Implementation of the Semiclassical Quantum Fourier Transform in a Scalable System." *Science* 308 (13 May 2005): 997–1,000. The researchers demonstrate a small-scale version of a quantum computer that could perform steps of quantum algorithms.

Churchhouse, R. F. *Codes and Ciphers: Julius Caesar, the Enigma, and the Internet.* Cambridge: Cambridge University Press, 2001. From ancient Rome to the digital age, cryptology includes a broad range of codes and ciphers. Churchhouse describes how cryptology has evolved and explains how the different codes and ciphers work.

Clark, Josh. "How Quantum Cryptology Works." Available online. URL: science.howstuffworks.com/quantum-cryptology.htm. Accessed June 5, 2009. In this article, Clark describes the differences between traditional cryptology and the techniques that will become available if and when quantum computers are developed.

Curtin, Matt. *Brute Force: Cracking the Data Encryption Standard.* New York: Copernicus Books, 2005. Some computer specialists believed that the Data Encryption Standard, adopted in the 1970s, was too weak to provide adequate encryption. This book tells the story of the people who demonstrated the weakness in the 1990s by breaking the encryption.

Ekert, Artur, Carolina Moura Alves, and Ajay Gopinathan. "History of Cryptography." Available online. URL: cam.qubit.org/articles/crypto/intro.php. Accessed June 5, 2009. The authors, at the University of Cambridge's Centre for Quantum Computation, provide a brief and well-illustrated history of cryptography and cryptanalysis.

Federal Trade Commission. "How Not to Get Hooked by a 'Phishing' Scam." Available online. URL: www.ftc.gov/bcp/edu/pubs/consumer/

alerts/alt127.shtm. Accessed June 5, 2009. The Federal Trade Commission, the United States government agency that regulates trade and commerce, offers tips to avoid phishing scams.

Krystek, Lee. "Cryptology: The Science of Secret Codes and Ciphers." Available online. URL: unmuseum.mus.pa.us/cipher.htm. Accessed June 5, 2009. Hosted by the Museum of UnNatural Mystery, this Web resource explains the basics of codes, ciphers, and cryptanalysis.

Network Associates, Inc. "How PGP Works." Available online. URL: www.pgpi.org/doc/pgpintro/. Accessed June 5, 2009. This introduction to the encryption program Pretty Good Privacy (PGP) explains public key cryptography, digital signatures and certificates, and how PGP implements these functions.

Niven, Ivan. *Mathematics of Choice: How to Count without Counting.* Washington, D.C.: The Mathematical Association of America, 1965. Niven lucidly explains combinatorial mathematics.

Ritter, Terry. "Learning About Cryptography." Available online. URL: www.ciphersbyritter.com/LEARNING.HTM. Accessed June 5, 2009. This cryptography tutorial takes the reader step-by-step through basic encryption techniques, including the strengths and limitations of each method.

Singh, Simon. *The Code Book.* New York: Anchor Books, 1999. Not only does this book provide easily understood descriptions of codes and ciphers, it also tells the stories behind the encryption methods—and how these methods were used, and sometimes broken, altering the course of history. A chapter on quantum cryptography is included.

———. *The Code Book: How to Make It, Break It, Hack It, Crack It.* New York: Delacorte Press, 2001. This book contains most of the material of Singh's 1999 version, but aimed at young readers.

Stripp, Alan. "How the Enigma Works." Available online. URL: www. pbs.org/wgbh/nova/decoding/enigma.html. Accessed June 5, 2009. The Public Broadcasting Service and the television series *Nova* sponsor this tutorial on the Enigma machine. Links to related pages, such as a program that simulates an Enigma machine, are also provided.

Sun Microsystems, Inc. "Introduction to Public-Key Cryptography." Available online. URL: docs.sun.com/source/816-6154-10/contents. htm. Accessed June 5, 2009. The computer company Sun Microsystems provides this document, which explains public key cryptography, digital signatures, certificates and authentication, and the Secure Sockets Layer (SSL) Protocol.

Web Site

National Security Agency. Available online. URL: www.nsa.gov. Accessed June 5, 2009. The National Security Agency performs cryptography and cryptanalysis in support of the U.S. government's need to protect sensitive information and collect intelligence on potential threats. On its Web site, the agency describes its history, mission, and cryptology research.

5

Computer Vision—Creating Machines That Can See

Building a computer that can mimic certain aspects of human intelligence and capability is a major goal of computer science. One of the most important of these capabilities is vision—the ability to see.

The value of visual information has long been recognized. To convince advertisers of the need for visual displays, advertising executive Frederick Barnard wrote in 1921 that a look was worth 1,000 words. Barnard's idea was that a picture included in an advertisement on a billboard or in a magazine would make a much more effective sales pitch than just words. Although nobody knows the exact number of words that a given picture can convey, the saying "a picture is worth a thousand words" has become a popular way of emphasizing the importance of images and illustrations in human thinking processes. The value of illustrations is demonstrated in their use throughout many science textbooks. Some concepts, particularly those related to geometry and spatial arrangements, are much easier to grasp with a picture rather than a verbal description.

A lot of information is available through vision, as indicated by the large portion of the human brain devoted to this sense. The cerebral cortex is a critical part of the brain that performs complicated functions such as memory, learning, and perception. Vision is so important that about half of the area of the cerebral cortex is involved in the processing of visual information. This is also true of monkeys, apes, and other primates, which rely on their visual sense to move around in their environment, find food, and avoid predators.

Engineers have long wanted to build machines capable of automatically navigating complex environments, which requires some sort of machine vision. Robots, first popularized in Czechoslovakian writer Karel Čapek's 1921 play *R.U.R.* (which stands for Rossum's Universal Robots), could accomplish a wide variety of essential and perhaps dangerous tasks, such as locating and defusing unexploded bombs, if only they could "see."

But vision, which comes naturally to humans, has been challenging to implement in machines, even the most advanced computers. Pictures are worth a lot of words, and due to this richness of information, images and visual scenes are difficult for a robot to understand—machines

Researchers at the National Institute of Standards and Technology and the Department of Homeland Security test the capability of this search-and-rescue robot to cross a pile of rubble. *(National Institute of Standards and Technology)*

tend to get fooled by shadows and glare, have trouble recognizing objects, and are sometimes unable to focus on the important aspects of their visual environment while ignoring the rest. Despite the difficulties, the creation of increasingly sophisticated computers has sparked the development of computer vision. This chapter describes research at the frontiers of computer science involving machines that recognize human faces, space probes that navigate their way around rocks on the surface of other planets, and cars that drive themselves.

INTRODUCTION

Artificial intelligence (AI) is a common theme in modern research in computer science. A component of many AI systems is autonomous maneuverability—the capacity to move and navigate without the aid of human guidance. Remotely controlled vehicles are maneuverable but not in the same way, since they require the eyes and hands of a human operator for steering. A robot capable of independently performing sophisticated jobs will need to sense its environment, and vision is one of the most crucial senses.

One way of equipping robots with vision is to mimic human vision. Biology has often inspired computer technology, such as the neural networks of the brain that guide researchers who design artificial neural networks.

In humans and many animals, vision begins when the cornea (which is the clear outer surface of the eye) and a lens made of deformable tissue focus light rays on the retina at the back of the eye. The retina contains tiny cells called photoreceptors that transform light energy into patterns of electrical activity. A human eye has about 125 million photoreceptors. The electrical activity is transmitted to other cells, usually by the release of certain chemicals called neurotransmitters, and the information travels along networks of neurons, as described in chapter 2.

Neural networks process information by extracting certain features such as contours, colors, or movement. Scientists have learned a little bit about how these networks operate by recording their electrical activity and relating this activity to a visual stimulus—an image or object presented to the eyes. For instance, when a person watches a basketball game, the activity of neurons in certain parts of his cerebral cortex identifies the important objects—players, the basketball, and the hoop—and separates them from the background. Neurons in other parts of the brain process color, such as the red or blue of a jersey. The activity of yet other neurons concentrates on motion, giving the viewer a perception of the movement of the basketball and players.

None of these tasks are trivial, though the human brain is so complex and finely tuned that vision seems to be effortless. Neural networks involved in vision specialize in object recognition, color perception, or motion detection, increasing overall efficiency. Human brains are so adept at interpreting contours that even a crude two-dimensional drawing done in pencil on a piece of paper can be recognizable.

Specialization is evident when parts of the brain become damaged, reducing or even eliminating the functions of this area. In some patients, an injury to a specific portion of the brain known as the fusiform gyrus results in prosopagnosia, a medical condition in which the ability to recognize faces is severely impaired. The patient can still see and recognize most objects, but has trouble recognizing the faces of friends or even family members.

Although scientists have made progress in the study of the brain, there is much about this astonishingly complex organ that eludes understanding. Engineers and computer scientists have built artificial neural networks capturing certain aspects of brain functioning, but complicated functions such as vision, which encompass so many areas of the brain, are not yet possible to implement in silicon and electronics.

It is not essential for a robot or machine to mimic human vision, particularly if there are simpler methods. The important requirement is the ability to form an image and extract the necessary information. Frenchmen Louis-Jacques-Mandé Daguerre (1789–1851) and Joseph Nicéphore Niépce (1765–1833) invented photography in the 1830s, creating a picture by using a lens to form an image on a plate containing light-sensitive materials. These early cameras gradually improved in resolution—the ability to distinguish small objects—and visual equipment such as movie cameras evolved. This technology can be the "eyes" of a robot or computer, forming images and inputting the data for further processing.

Interpreting this visual data is not so easy. Programming computers to do repetitive functions such as arithmetic or searching through a list is simple, but programmers are unsure how to design algorithms to extract information from pictures. Early devices, such as the simple electronic robots designed and built in the 1940s by neuroscientist

William Grey Walter (1910–77) at the Burden Neurological Institute in Bristol, England, had sensors that could react to the presence or absence of light, but had little or no visual capacity.

One of the earliest autonomous robots was Shakey, built and studied at SRI International (then known as the Stanford Research Institute) in California during the years 1966 through 1972. Shakey, whose name came from its jerky movements, was six feet (1.83 m) tall with a television camera and antenna mounted on top and wheels attached to motors on the bottom. Because computers were so big in those days Shakey did not have a computer on board; instead, the robot used a radio link to communicate with a stationary computer. The camera produced images, which the transmitter relayed to a computer known as a PDP-10. Programs running on this computer controlled the robot's motion. In July 2004, Carnegie Mellon University in Pittsburgh, Pennsylvania, inducted Shakey and other early robots into the Robot Hall of Fame. James Morris, a professor of Computer Science at Carnegie Mellon University, said in an SRI news release issued July 12, 2004, "SRI's Shakey was a true pioneer, showing that truly autonomous robotic behavior was feasible long before anyone else."

Although Shakey was capable of independent motion, the robot could only maneuver in a special environment. To simplify the problem of visual analysis, Shakey's developers, led by Charles Rosen, carefully designed an environment that presented only simple geometrical shapes, illuminated by special lighting systems to reduce shadows. The robot's algorithms were able to extract boundaries and corners from the visual scene because these features were clearly presented and of a limited number of orientations. Shakey would have been hopelessly lost in a complex environment outdoors.

Improvements in computers and algorithm efficiency led to the modern digital age. These improvements resulted in faster and better image processing tools, and spurred the research and development of more complicated computer vision systems.

IMAGE PROCESSING AND DIGITAL IMAGES

Cameras and film, using techniques that date back to Daguerre, are still common today, but the newer techniques of digital photogra-

Sensor for a digital video camera *(GIPhotoStock/Photo Researchers, Inc.)*

phy are becoming increasingly popular. A digital image consists of an array of picture elements. Each picture element, or pixel, represents the brightness and/or the color of a small area of the picture with numbers.

Whereas photographic film uses reactions with light to reproduce a visual scene, digital cameras have light-sensitive electronic sensors to detect light. A lens focuses the light, as in a camera that uses film, but the sensor of a digital camera contains many pixels to convert the image into an array of values representing the amount of light and its color. The number of pixels varies, depending on the quality of the camera, but most digital cameras have millions of pixels. These values are stored in the digital camera's memory, and can be downloaded onto a computer or printed with a device that converts the digital image into a viewable picture—the analog representation. Electronic equipment called scanners reverse this process, converting analog photographs into digital format. Any kind of signal can be turned into a digital one. The figure on page 128 illustrates the process of making digital representations.

Digital images are usually less subject to noise—unwanted signals—than analog images. Another major advantage of digital images

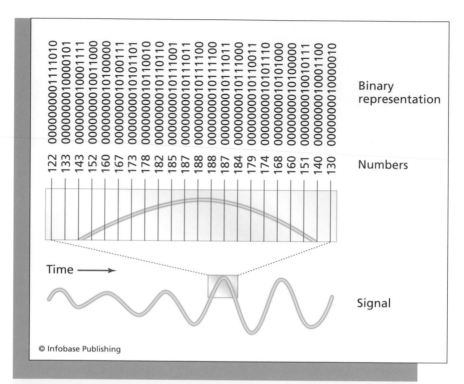

The signal to be converted into digital format appears at the bottom. The digitizing process, shown for the highlighted portion of the signal, converts the values of the signal into binary representations.

is that they can be processed with computers. But the data of even one picture can be huge, especially if there are a lot of pixels, which take up a lot of memory. To avoid filling up too much memory, special algorithms compress the data without losing too much information. These compression algorithms take advantage of situations such as a broad, uniform area—a solid white wall, for example—that can be encoded in a simple manner rather than being represented by a large number of pixels, as it is in the original or "raw" image. One of the most common compression techniques is *JPEG*, which stands for Joint Photographic Experts Group, the name of the committee that developed it. JPEG images take up much less space in memory or when stored in a computer file on a hard disk.

Compression is a simple type of image processing. Other image-processing algorithms can be much more complicated, with sophisticated computer programs capable of making insertions and deletions to alter an image, combine images, or introduce a new object in the image that was not present in the original or remove one that was. Images in certain movies and video games can be terrifically realistic with these techniques.

Although image alteration is beneficial, it has its dark side as well. Analog photographs can be modified by techniques such as airbrushing, in which an air-operated dispenser emits a fine spray of paint, but these techniques are crude and generally easy to spot when compared to digital image processing. The output of a digital camera, or an image digitized by a scanner, is an array of numbers, which in binary format is perfect for computers. Computer processing of images allows manipulation that is so effective and realistic that it is difficult to detect. As a result, images and photographs can be forged with the intention of embarrassing, defaming, or defrauding people.

But computer scientists such as Jessica Fridrich at Binghamton University, of the State University of New York, are developing countermeasures. Digital cameras generate patterns in the numbers composing the image, even though these patterns may be undetectable when a person examines the picture in analog format. The pattern is specific to a camera, somewhat like the fingerprint of a person. In 2006, Fridrich and her colleagues searched for these patterns by examining several hundred images taken with nine different digital cameras; using the patterns, they were able to identify the camera that had taken each image. Jan Lukáš, Fridrich, and Miroslav Goljan published their report, "Digital Camera Identification from Sensor Pattern Noise," in a 2006 issue of *IEEE Transactions on Information Security and Forensics*.

Matching an image with the digital camera that took it is often important in authenticating a picture. In their report, the researchers wrote, "Reliable identification of the device used to acquire a particular digital image would especially prove useful in the court for establishing the origin of images presented as evidence. In the same manner as bullet scratches allow forensic examiners to match a bullet to a particular barrel with reliability high enough to be accepted in courts, a digital equivalent of bullet scratches should allow reliable matching of a digital image to a sensor."

This technique can also be important for detecting frauds. An alteration in the image changes the pattern, which might permit an image expert to discover the manipulation.

Techniques to modify or authenticate digital images are sophisticated ways of processing images. What robotics researchers would like to do is to find processing techniques to analyze images and extract features such as obstacles that an autonomous robot must avoid. These techniques would allow a robot to "see" and interpret images.

Useful visual systems must be effective at spotting patterns. These patterns are not necessarily the same as those discovered by Fridrich and her colleagues, but are instead visual representations of an object's geometry, such as the round pattern made by the boundary of a basketball, the outline of a tree, and so forth. The human eye and brain are excellent pattern detectors.

One method of achieving pattern recognition in machines is to employ methods similar to those of the brain. Artificial neural networks are a popular means of implementing brainlike technology, and computers can be used to simulate a variety of neural mechanisms.

Thomas Serre, Aude Oliva, and Tomaso Poggio of the Massachusetts Institute of Technology (MIT) recently designed a computer model to simulate early visual processing occurring in the human brain. Humans can scan a visual scene and rapidly categorize objects—a person needs little time to distinguish a horse from a car. In the human visual system, information travels in the "feedforward" direction, that is, from the early processing stages to advanced processing centers in the cerebral cortex—as well as going in the reverse, "feedback" direction, in which the advanced processing centers adjust or guide the output of the early processing stages. The quickness of categorization implies to researchers such as Poggio that there is little time for feedback information flow to affect this process, so the model of Poggio and his colleagues incorporates only the feedforward path. Their model, as described in "A Feedforward Architecture Accounts for Rapid Categorization," published in a 2007 issue of the *Proceedings of the National Academy of Sciences,* displays behavior and object categorization similar to that of a person.

But computer vision need not work on the same principles. Although there are similarities in the way the brain processes information and the way a computer acts on data, the brain relies on the electricity activity of neurons, while the computer operates on binary data—1s and 0s—controlled by digital logic circuits. Perhaps the best way to get a computer to "see" patterns and recognize objects, at least until biologically inspired technology becomes sufficiently advanced, is to use the speed of computer processors and sophisticated mathematical algorithms.

OBJECT RECOGNITION

Visual images are rich in information, as indicated by the old adage that a picture is worth a thousand words. An algorithm to extract patterns or features from this information-rich source usually strives to minimize the amount of information it has to slog through in order to find what it is looking for. In other words, to find a needle in a haystack more efficiently, reduce the size of the haystack. One way of accomplishing this is to reduce the dimensions—the parameters or measurements that describe an object or system of objects.

Consider a flat sheet of paper. This is a two-dimensional object, having a left-right axis and an up-down axis, but no depth if it is considered to be completely flat. The paper can be described using only two dimensions.

Now consider what happens if someone crunches or folds the piece of paper into a certain shape. The shape has three dimensions because it has depth. A mathematical description of the shape, such as a listing of the coordinates of each point, will need to employ three-dimensional space. But suppose a researcher is only interested in the paper. The paper is still a connected object, and although it is no longer in a plane, the paper itself has only two dimensions. Mathematicians have developed complicated procedures known as transformations to convert higher-dimensional descriptions, such as the shape of the paper, to lower ones, such as the two-dimensional paper.

Such transformations are even more important in other problems. Space consists of three dimensions, but mathematical objects can have a lot of dimensions, depending on the number of variables needed to describe them. Transformations provide much needed simplifications. Computer programmers can use these transformations and other mathematical operations to extract features from complicated scenes. If the feature is a boundary of an object, such as the outline of a box or an airplane, the object can be recognized.

The problem is that it is not obvious what transformation or which algorithm will work for any given scene. Scenes with simple geometrical objects, such as the environment of Shakey the robot, present few difficulties. But typical visual scenes involve curves such as winding roads or objects with twisting, complicated outlines. Some algorithms may succeed for one scene but fail for another.

But researchers such as Aleix Martínez at Ohio State University have studied which algorithms work best in a particular application. Martínez and his student Manil Zhu tested algorithms of a certain kind, called linear feature extraction methods, on two different types of problem—sorting different objects such as apples and tomatoes, and identifying facial expressions. The researchers developed a test that predicted an algorithm's success or failure in these different tasks. Applying a test in advance saves a researcher time by guiding the selection of the most promising algorithm, instead of having to pick a random algorithm that may or may not work. Martínez and Zhu reported their results in "Where Are Linear Feature Extraction Methods Applicable?" published in 2005 in *IEEE Transactions on Pattern Analysis and Machine Intelligence*. The researchers believe their work will help computer vision researchers "to better understand their algorithms and, most importantly, improve upon them."

Researchers who test algorithms, as do Martínez and Zhu, and those who create or develop them often need some sort of standard of measurement. But if researchers use different sets of pictures sometimes their results will vary, depending on which set of pictures was used. To standardize the process, researchers often use sets, called databases, of pictures or elements that have already been studied. By using such standard databases, researchers can compare their results without worrying about differences created when different sets of pictures are used. For example, Martínez and Zhu used a database called ETH-80, a set of 80 pictures of objects, which was developed by scientists at Darmstadt University of Technology in Germany.

Another important database is Caltech 101. This large database contains pictures of 101 categories of objects, with each category containing 40 to 800 images. The pictures show a large variety of natural objects, and researchers frequently use this database to measure the success rate of their object-recognition algorithms or computer programs. If an algorithm can successfully recognize many of the objects represented in this database, researchers believe the algorithm will be successful in the real world.

By testing algorithms with this standard, and tweaking the algorithms if necessary, computer scientists have developed computer vision systems that successfully recognize as many as two-thirds of the objects. Although not perfect, these success rates are encouraging—and give Shakey's successors a chance of navigating in various outdoor terrains without rolling into a ditch or falling over a log.

Object-recognition algorithms are also useful in situations such as traffic control. There are not enough police officers to be present at all places and at all times, so many cities have started monitoring some of their busier streets and intersections with camera systems equipped with the technology to measure vehicle speed or position. This technology is able to detect traffic violators, and with the aid of object-recognition systems, observation and monitoring is usually automated. As described in the following sidebar, cameras take a snapshot of the violator's car, and computer vision algorithms locate and read the license plate. After looking up the registered owner's address, the police mail a ticket to the offender.

Since computer vision is not perfect, law enforcement officials supervise and constantly check license plate recognition systems involved in traffic ticketing. Although success rates are high in many object recognition tasks, computer vision algorithms make mistakes. One of the biggest reasons for these mistakes is not due to poor feature extraction but a failure to apply context. An object may look similar to another object in many cases, but humans get clues about an object's identity from considering the context in which it is found. For instance, a certain object may appear to be a muddy watermelon, but if it is sitting on a field marked with stripes, and surrounded by players wearing shoulder pads and helmets, the object is probably a football. Lacking these contextual clues, a computer often goes astray.

Some researchers, such as James DiCarlo at MIT, have criticized other computer vision researchers for relying too much on databases like Caltech 101. A standard set of images makes it possible to compare algorithms, but provides only a limited number of perspectives and contexts. With colleagues Nicolas Pinto and David D. Cox, DiCarlo published a report, "Why Is Real-World Visual Object Recognition Hard?" in a 2008 issue of *Public Library of Science Computational Biology*. The researchers argue that "while the Caltech 101 set certainly contains a large number of images (9,144 images), variations in object view,

Getting a Traffic Ticket in the Mail—License Plate Recognition

Reading a clean and visible license plate is easy for people with good eyesight. Computer algorithms to read text, such as optical character recognition systems described in chapter 2, are also effective if the characters are printed in a standard font. But the problem of reading a license plate is not so simple. First, the plate must be found. Although this task may sound simple, it is not so easy for a computer. A license plate has a general format and location, but can vary slightly from state to state and car to car.

License plate recognition systems often have an infrared camera, which can take photographs in any level of light, including at night, since all warm objects emit a certain amount of infrared radiation. Monitoring systems, such as radar to detect a vehicle's speed and position sensors to specify its location, trigger a photograph when a car exceeds the speed limit or runs a red light. Computer vision algorithms search the picture for a small rectangular object—the license plate. Once found, feature extractors pull out the largest figures, which will be the letters and numbers of the license plate.

position, size, etc., between and within object category are poorly defined and are not varied systematically." These limits make the problem of object recognition simpler than it really is, artificially inflating the success rate of computer vision algorithms.

To illustrate their point, DiCarlo and colleagues built a simple model based on brain structures. This model performs as well as the best algorithms at identifying objects in standard databases, yet fails dismally when tested in natural environments. DiCarlo and colleagues argue that computer vision researchers have been sharpening their al-

Software to read these characters must be capable of identifying a variety of sizes and fonts; many of these programs employ neural network programs. Other information, such as the issuing state, may also need to be read. When finished, the computer searches the registration database for the owner's name and address. The violator will receive a ticket in the mail.

Traffic monitoring systems such as red-light cameras have been used for several decades, but their popularity has increased in the past 10 years. Some people protest against what they believe is a violation of their privacy, but according to a 2002 Gallup poll, 75 percent of motorists approve of these devices. These motorists realize that such devices make the roads safer. For example, in a study of red-light cameras installed at two major intersections in Philadelphia, Pennsylvania, researchers found that red-light cameras reduced red-light running by more than 90 percent. The researchers, Richard A. Retting and Charles M. Farmer of the Insurance Institute for Highway Safety, and Susan A. Ferguson of Ferguson International, published their findings in "Reducing Red Light Running through Longer Yellow Signal Timing and Red Light Camera Enforcement: Results of a Field Investigation," in a 2008 issue of *Accident Analysis & Prevention*.

gorithms on simplistic images rather than in complex, real-world environments, and as a result, computer vision is not making much progress in solving navigation or other problems. In their report, DiCarlo and colleagues conclude that their "results demonstrate that tests based on uncontrolled natural images [such as databases like Caltech 101] can be seriously misleading, potentially guiding progress in the wrong direction."

The issue of database usage is important to study further. As computer vision tackles jobs of increasing difficulty—and importance—

success will be harder to achieve. This is particularly true for the critical task of recognizing faces.

MATCHING NAMES WITH FACES—FACIAL RECOGNITION SYSTEMS

Imagine a terrorist who gets on a plane and enters the United States. The terrorist breezes through security with a fake passport and other falsified documentation, but upon exiting the terminal, a swarm of police descends and arrests him. Police in this scenario did not receive a tip from an informant, but instead received a tip from camera surveillance, along with a computer vision system that recognizes faces. Unbeknownst to the terrorist, his face and identity had been collected by law enforcement authorities around the world, and the computer alerted police in the United States the moment he stepped off the plane.

Similar to red-light cameras, surveillance systems maintain a law enforcement presence 24 hours a day. Surveillance such as continuously running cameras in banks and stores have helped convict many thieves and robbers, but in the case of a terrorist who is planning on carrying out widespread destruction, it is imperative to stop the crime before it is committed.

The difficulties in implementing the hypothetical system described above are numerous. A facial recognition system must deal with the variety of angles and lighting in which a face may be seen, the changes due to aging, and different facial expressions that distort or alter the face, as well as a prepared terrorist who has adopted a disguise. This is a "real-world" problem with no simple solution.

How can a computer vision algorithm possibly succeed? One way is to note the characteristics that make individual faces unique. This procedure is similar to the old Bertillon system, invented by Frenchman Alphonse Bertillon (1853–1914) in the 1880s to identify criminals. In this era, before the development of fingerprinting technology, the justice system had no certain means of identifying suspects and perpetrators. Bertillon made a careful study of a number of anatomical measurements, such as foot length, finger length, skull width, eye color, and others, and chose a set of measurements that are unique to an individual. By measuring all arrested persons and suspects, the police maintained a file that could be consulted to check if the person had been arrested earlier and for what crime.

Facial recognition systems can use distinguishable traits or features of a face that do not change much over time and are difficult to alter or disguise. A human face has about 80 characteristic features, known to computer scientists as nodal points. These nodal points include the distance between the eyes, the contours of the cheekbones, shape of the nose, and depth of the eye sockets. Facial recognition algorithms compute these points from an image and search the database to match a name to the face, if the face has already been recorded. Not all of the 80 features need be measured—individual faces may share the same measurements for a few of these nodal points, but 15 or 20 points are generally enough to distinguish any two faces.

Engineers and computer scientists have begun to develop facial recognition algorithms. Las Vegas casinos use facial recognition software in the attempt to identify gamblers in their casino who are in databases of known or suspected cheaters. Law enforcement authorities tested a facial recognition program while it scanned attendees of Super Bowl

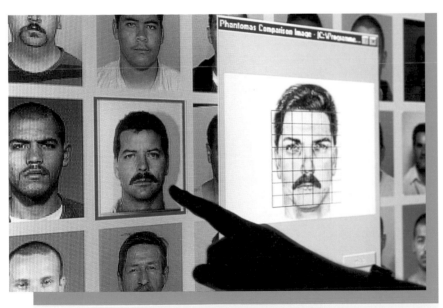

Face recognition systems analyze characteristic features to find a match. *(Thomas Pflaum/VISUM/The Image Works)*

XXXV in Tampa, Florida, on January 28, 2001. In 2008, the Oregon Department of Motor Vehicles began using facial recognition software to prevent someone from obtaining a license or identification card under an assumed name.

But facial recognition algorithms, like other object recognition systems, have not yet been perfected. Facial recognition software used today requires constant monitoring and human input to reduce mistakes. The potential benefits of a fully functional and automated facial recognition system have motivated the government and other interested parties to sponsor evaluations of current systems. An evaluation called Face Recognition Vendor Test (FRVT), designed to evaluate commercially available systems, has taken place periodically since 2000. FRVT 2006 received systems from 22 companies and universities for testing. Testing is not a sim-

United States Department of Homeland Security

The terrorist attacks on American soil on September 11, 2001, in which terrorists hijacked planes and destroyed the World Trade Center towers in New York City, horrified the nation. One of the responses to these attacks was the creation of the Department of Homeland Security, established by the Homeland Security Act of November 25, 2002. The new department assumed operation of functions such as immigration services, border patrol, the Coast Guard, disaster preparedness and response, the Secret Service, and others. It was the biggest restructuring of federal government organization in five decades.

The 2008 budget of the Department of Homeland Security was nearly 50 billion dollars. Included in its operations is the Directorate for Science and Technology, the branch of the department devoted to research and development projects. To improve homeland security, the department

ple process, and involves a broad range of applications and procedures, since the use of only one or two standards or databases can sometimes be misleading. The National Institute of Standards and Technology (NIST), the United States government agency devoted to improving the science and technology of measurement, conducted FRVT 2006.

Several U.S. government agencies sponsor these tests. One of the most prominent is the Department of Homeland Security. As described in the following sidebar, this agency is responsible for protecting the United States against terrorist attacks. In the war against terror, homeland security officials have mustered all available surveillance tools and are eager to develop more, including facial recognition systems.

Obstacles yet to be surmounted in facial recognition systems include the problems of lighting, shadow, and the variety of perspectives

provides support and funding to such projects as the Face Recognition Vendor Test along with other technological advancements such as more efficient screening of airport baggage and methods to detect nuclear devices and other highly destructive weapons. The Department of Homeland Security maintains several government laboratories, such as the National Biodefense Analysis and Countermeasures Center at Fort Detrick, Maryland, and the Transportation Security Laboratory in Atlantic City, New Jersey. Funding is also provided to researchers at universities and other institutions for special projects.

One of the main goals of the department is the prevention of attacks. To counter the threat posed by terrorists entering the country, the Department of Homeland Security monitors and verifies visitors with established technologies such as digitally scanned fingerprinting. Department officials are also investing in the establishment of databases containing faces of terrorists and their associates, along with facial recognition systems to automate and extend police screening at airports and other entry points into the country.

and angles from which a face can be viewed. These difficulties are shared by all object recognition systems.

Another problem involves facial expressions—faces are not etched in stone, but can smile and frown. Since smiles can thwart facial recognition technology, passport authorities in many countries, including the United States, will not accept a photograph from a passport applicant with a "toothy grin."

These problems have led to failures in early versions of systems when tried in real-world situations. In September 2003, the *Boston Globe* and other news organizations reported that a facial recognition system tested at Logan International Airport in Boston, Massachusetts, failed to recognize a large percentage of test cases.

But computer scientists are continuing to refine the technology. Rob Jenkins and A. Mike Burton at the University of Glasgow, in Scotland, have found a way to improve recognition accuracy by averaging across a number of photographs. The researchers published their report, "100% Accuracy in Automatic Face Recognition," in a 2008 issue of *Science*. Variability associated with facial expressions, viewing angles, and changes due to aging are difficult for current technology, but as Jenkins and Burton discovered, "Averaging together multiple photographs of the same person dilutes these transients while preserving aspects of the image that are consistent across photos. The resulting images capture the visual essence of an individual's face and elevate machine performance to the standard of familiar face recognition in humans." Computer algorithms were much more likely to recognize a composite image of a face obtained by merging about 20 different photographs.

Accurate facial recognition systems would aid police investigations and enhance homeland security, but such systems also raise privacy concerns. Public surveillance and facial recognition technology might allow government officials to track and spy on citizens.

But these systems have many potential benefits, including the possibility of offering citizens a stronger degree of identity protection. University of Houston researcher Ioannis Kakadiaris and colleagues are developing recognition systems to provide a unique identity, similar to fingerprints, which would be useful in authorizing credit card purchases and other transactions. The system designed by Kakadiaris performed extremely well in FRVT. If businesses eventually adopt this system, proving one's identity may be as simple as posing in front of a camera or webcam.

AUTONOMOUS VEHICLES—MACHINES THAT DRIVE THEMSELVES

Recognizing and identifying faces is important for homeland security and police investigations, but other computer vision applications require different proficiencies. One of the main goals of computer vision since the development of Shakey the robot has been navigation and maneuverability. Although some applications can use human eyes and brains—an operator controls the motion of a vehicle from a remote location via radio transmissions—there are applications in which remote control is not feasible.

One of these applications is the control of machines that explore other planets. For example, radio communication with a vehicle on Mars, which at its closest is about 35,000,000 miles (56,000,000 km) away from Earth, would be far too slow to let a human operator on Earth control the vehicle. At this distance, radio takes more than three minutes to make the trip—the vehicle could smash into a rock long before the arrival of the human operator's signal to stop!

The U.S. agency devoted to space exploration and research, the National Aeronautics and Space Administration (NASA), as well as similar agencies in other countries, build and launch probes to travel throughout the solar system. Some of these unmanned probes land on other worlds, such as Mars. There are obviously no roads or smooth trails on these worlds. Terrain on most of Mars, for instance, is sandy and rocky. Yet exploring the ground with a mobile vehicle offers vastly more information than a stationary probe that is stuck in one spot—imagine sampling a whole planet at only one point!

Until manned vehicles to Mars are possible, unmanned probes must suffice. Since these exploratory probes cost millions of dollars to build and launch, NASA needs an accurate guidance system to avoid disaster.

Mars has captured the fascination of many people because of its proximity and its similarity to Earth. Present conditions on the planet are harsh, and the atmosphere is extremely thin, but surface features resembling water channels, as seen by orbiting probes, suggest that Mars may have once been warm enough to have flowing water. And in 1996, NASA scientist David McKay and his colleagues reported that they had found evidence of primitive fossil life in a meteorite that came from Mars. (The meteorite was found in Antarctica, having been ejected by some natural process from Mars, and later landed on Earth.) Although this finding is controversial, people have long speculated on the possibility that life evolved on Mars as it did on Earth. But proving the existence of life, or a lack thereof, requires more substantial evidence.

In 1976, NASA landed two probes, called Viking I and Viking II, on Mars. One of the main goals of these missions was to conduct chemical and biological tests on soil, which a robotic arm scooped from the ground. These probes were stationary, so they sampled from only two places—the landing spot of each probe. Tests were inconclusive, showing some activity that suggested life, though the results could also be easily explained by nonbiological chemical reactions.

More recent NASA probes to Mars have been mobile. *Pathfinder,* launched on December 4, 1996, arrived on Mars on July 4, 1997, and included a rover called Sojourner that traveled short distances. This rover was about the size of a microwave oven.

The Mars Exploration Rovers are a more ambitious mission, with *Spirit* launching on June 10, 2003, and *Opportunity* launching on July 7, 2003. These rovers are slightly smaller than a golf cart, standing about 4.9

Mars Exploration Rover *(NASA)*

feet (1.5 m) high, 7.5 feet (2.3 m) wide, and 5.2 feet (1.6 m) long. Powered by solar energy, they were designed to roam the surface of Mars for up to about 0.6 miles (1 km). *Spirit* landed on January 4, 2004, and *Opportunity* on January 25, 2004. (Both landing dates are Universal Time, a precise time standard.) Each rover houses 20 cameras, some of which are used for scientific purposes and some for navigation. The rovers were not only successful, they far exceeded expectations—as of May 27, 2009, *Spirit* had traveled a total of 4.8 miles (7.73 km)—although it is presently stuck in soft terrain—and *Opportunity* about 10.1 miles (16.2 km).

Human operators on Earth selected the routes for some of these journeys. Mission specialists examined the pictures transmitted from the rover cameras, programmed the routes, and sent the appropriate commands to the motor.

In other cases, the rover had to choose the path itself. These vehicles do not travel very quickly—about five feet (1.5 m) per minute, and even slower when they are doing their own driving—but they could easily get toppled by a rock or stuck in a hole. If this occurred, the mission would probably be over.

Without a human operator at the controls, how do the rovers steer themselves? *Spirit* and *Opportunity* are bigger and more mobile than Sojourner, and require a more sophisticated automatic navigation system. The twin rovers use a pair of cameras to provide a stereoscopic three-dimensional view, similar to the three-dimensional perspective given by a person's two eyes. As described in the following sidebar on page 146, navigation software running on a computer installed on the rover locates dangerous obstacles and maneuvers the vehicle to avoid them. NASA engineers and scientists developed this software based on algorithms discovered by researchers at Carnegie Mellon University.

The technology of the Mars Exploratory Rovers mission has performed exceptionally well, and scientists have learned much about the planet. Extended journeys have allowed scientists to sample a larger area of the surface than ever before. Chemical composition and the layering of certain rocks, studied with the rovers' sensitive scientific instruments, confirm that water once flowed on the surface of Mars.

Computer scientists and NASA engineers will continue to improve autonomous navigation algorithms for use in future unmanned missions. But computer vision also has an important and similar application here on Earth—driverless vehicles.

Remote operators can control unmanned vehicles on Earth much easier than on another planet, since radio communications are nearly instantaneous at short range. The military operates unmanned reconnaissance or attack aircraft, such as the U.S. air force unmanned airplane Predator, and bomb disposal squads use remotely controlled robots to defuse dangerous ordnance. But truly autonomous vehicles would be capable of many more functions, freeing military personnel from having to perform a large number of hazardous tasks. Driverless passenger cars could also provide greater mobility for those people who are unable to drive, as well as relieving tired or distracted drivers.

But designing an autonomous car capable of scooting down a highway is more ambitious than it sounds, and the problems of autonomous steering for driverless cars are more difficult than those faced by the Mars rovers. For one thing, the speed of passenger vehicles is much greater; the speed of the Mars rovers, five feet (1.5 m) per minute, is about 0.06 miles (0.036 km) per hour, about 1,000 times slower than highway speeds on Earth. And although the roads on Earth are (generally) smoother and straighter than the rocky Mars terrain, there is much more traffic and many more potential hazards.

Consider what a computer vision system must accomplish in order to steer a car. The system must recognize a variety of different road surfaces, each having different colors and textures, under a variety of lighting conditions. It must follow the paved surface through all of its twists and curves, and navigate complex routes including intersections and on- and off-ramps. Potholes, road debris, and other vehicles must be recognized and avoided.

No existing computer vision algorithm can perform these tasks efficiently and safely. Yet driving a car is certainly possible—the human brain and body do it with ease (at least when the person is paying attention). Finding a computer algorithm to do it is important enough for the Defense Advanced Research Projects Agency (DARPA) to sponsor a series of challenges encouraging computer scientists and engineers to tackle the problem.

DARPA GRAND CHALLENGE

DARPA is a research and development organization of the U.S. Department of Defense. The organization manages a large number of research projects relating to national defense, and targets projects on the frontiers of science and technology that lead to significant breakthroughs. Computer vision and autonomous vehicles are among the top priorities.

To stimulate research on autonomous vehicles, DARPA sponsored three "challenges," in which research and development teams competed for prizes. Each of the challenges involved a course or track that an autonomous vehicle must successfully navigate, with a cash prize awarded to the winners. Research teams entered the competition by registering their vehicles, which had to be unmanned and have no means of remote control. These races were the first long-distance competitions for fully autonomous vehicles.

The first grand challenge, held on March 8–13, 2004, in the Mojave Desert in southern California and Nevada, showed how difficult the task would be. On the final day competitors had to complete a 142-mile (227-km) course from Barstow, California, to Primm, Nevada. The course included roads (which were cleared of traffic for the race) and off-road terrain, marked by boundaries beyond which the driverless vehicles could not stray or they would be disqualified. All types of surfaces and terrain were included—steep slopes, paved and unpaved roads, rocky trails, sand, and even shallow water. The prize was $1 million to the first team to complete the course within 10 hours. A total of 106 research teams entered the competition, and after preliminary evaluations, DARPA narrowed the field to the best 15 to test their vehicles in the race.

No one won the 2004 grand challenge. One by one, the contestants strayed too far from the course, suffered mechanical breakdowns, collided with obstacles, or got stuck. The farthest any of the vehicles got was 7.4 miles (11.8 km), accomplished by the Red Team from Carnegie Mellon University. The Red Team's vehicle, called *Sandstorm,* was a modified Humvee, a military vehicle also known as HMMWV (High Mobility Multipurpose Wheeled Vehicle). *Sandstorm* strayed off course on uneven terrain and bogged down.

Competitors fared better in the second grand challenge, also held in the Mojave Desert, on October 8, 2005, because they learned from their previous mistakes. Autonomous vehicles of five of the 23 competing teams completed a 131.2-mile (210-km) course from Barstow, California, to Primm, Nevada. The fastest time belonged to *Stanley,* a vehicle entered by Stanford University, which finished the course in six hours, 53

Self-Steering in the Mars Exploration Rovers

Sometimes NASA specialists program a route and transmit it to a rover, but sometimes the best route is not obvious to those sitting back on Earth. In these cases, NASA personnel select a destination and transmit the coordinates, letting the rover do the navigating.

To get a three-dimensional view of the terrain, the rover combines two views taken from a pair of cameras that are spaced slightly apart and pointing in the same direction. Each of the two views provides a slightly different perspective, as does each of a person's two eyes. Three-dimensional scenes have depth, which can be computed from parallax— the change in an object's position when viewed at a different angle, as illustrated in the figure. (Parallax can be observed by holding a finger in front of the face and alternately closing one eye and then the other. The finger seems to move relative to the background.) The brain calculates depth by analyzing information from the two eyes, and the rover computers do the same from images taken by the two cameras.

With a three-dimensional view, a rover can generate a three-dimensional map of the terrain it must traverse. The navigation software determines the location of obstacles such as rocks and holes, along with their height or depth. Algorithms compute a number of possible paths to reach the target destination safely and evaluate each one. After choosing the shortest path consistent with a safe movement, the rover travels about 1.6–6.6 feet (0.5–2.0 m), then stops and reevaluates the route. This process repeats until the rover reaches its goal or receives a command to stop.

Although the software was successful, the original algorithms were slow and limited—sometimes the rover

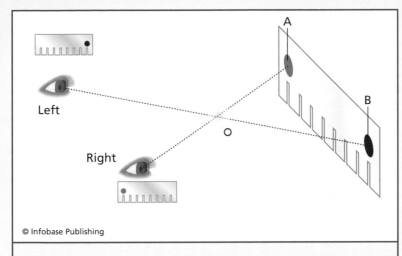

© Infobase Publishing

From the right eye's viewpoint, object O is seen at a position marked A. The left eye views the object from a different angle, and O appears at the position marked B.

could not find the best route because it could not plan far enough ahead, so it had to stick with simple routes. In summer 2006, NASA upgraded the navigation programs by transmitting new algorithms and loading them onto the computers. The new autonomous navigation programs, based on the algorithms of Tony Stenz and Dave Ferguson at Carnegie Mellon University, were originally designed for unmanned experimental combat robots for the U.S. army, but they have also been used in a number of mobile robots at Carnegie Mellon University. With the help of these algorithms, the Mars rovers have a more extensive "picture" of the terrain and are able to plot more complicated paths. If necessary, the rovers even have the capability to find their way out of a maze.

Vehicle participating in a DARPA challenge *(Defense Advanced Research Projects Agency [DARPA])*

minutes, and 58 seconds. In second place was *Sandstorm,* Carnegie Mellon University's Red Team entrant, at seven hours, four minutes, and 50 seconds. Third was another Red Team vehicle, a modified Hummer H1 called *H1ghlander* (pronounced "highlander"), about 10 minutes behind. The Stanford University researchers won $2 million for first place.

These research teams proved that computer vision algorithms are capable of maintaining a car on course while avoiding rocky obstacles on a variety of road surfaces and off-road terrain. But can these driverless vehicles maneuver in traffic? The final test, the third grand challenge, aimed to evaluate driverless vehicles in citylike conditions. Held on November 3, 2007, in Victorville, California, at the site of a closed air force base, this test was known as Urban Challenge. Out of a total of 89 entrants, DARPA officials selected the top 11 finalists to compete in the race. The competitors ran a 60-mile (96-km) course simulating congested traffic, with the autonomous vehicles required to obey traffic regulations and avoid hitting other vehicles. The winner was *Boss,* a robotic Chevrolet Tahoe modified by Carnegie Mellon University researchers, which finished the race in four hours, 10 minutes, and 20 seconds. Second place went to a robotic Volkswagen Passat named *Junior* from Stanford University, which trailed Boss by about 20 minutes. Close behind *Junior* was *Odin,* a Ford Escape Hybrid modified by Virginia Tech researchers. The Carnegie Mellon University team won $2 million, while the Stanford University researchers received $1 million, and Virginia Tech researchers took home half a million dollars.

Although the average speed was slow—Boss averaged about 14 miles (22.4 km) per hour—maneuvering in traffic required the computer vision algorithms to make quick decisions, just like any human driver. DARPA's challenges succeeded in sparking the development of computer systems that can provide autonomy under a variety of demanding conditions. With these improvements, along with continued research, computer vision may be ready for the highway before too long.

CONCLUSION

Computer vision is not yet as quick or discerning as human vision, but the progress in this frontier of computer science over the last few decades has been enormous. From a hesitant robot called Shakey, who was confined to environments containing only objects with simple geometry, researchers have gone on to develop autonomous vehicles capable of completing obstacle courses, maneuvering on Mars, and negotiating through traffic. The next step is to close the gap between computer vision and human vision.

Perhaps researchers will discover algorithms that closely mimic human vision, or perhaps researchers may construct brainlike mechanisms such as artificial neural networks, which will have visual capacity similar to humans. Another strategy is to design hybrid systems to exploit the advantages of both humans and computers.

One field of research aiming to meld human and computer systems involves work on prosthetics—artificial body parts. Visual prostheses would help people whose vision has been greatly impaired or even lost. Vision is an important sense in humans, but more than 1 million Americans have seriously impaired vision. Blindness often results from eye damage or diseases of the retina, although injuries to parts of the brain involved in processing visual information can also lead to visual impairment.

Mark Humayun, a researcher at the University of Southern California and the Doheny Retina Institute, is conducting efforts to develop an artificial retina. The device consists of an array of 16 electronic elements to convert light into electrical signals. These elements take the place of photoreceptors that have been damaged in a disease known as retinitis pigmentosa. After the device is implanted in the eye, it transmits signals electrically to neurons, which pass the signals along to visual areas in the brain. Humayun and his research team successfully implanted artificial retinas in six patients in 2002. The device has given these patients simple visual capacity, such as detecting light and distinguishing certain objects, but the limited number of elements cannot hope to provide the same quality of vision as the millions of photoreceptors in a normal retina. In 2007, Humayun increased the number of elements to 60, which will allow for greater visual acuity.

Other research projects have opted for similar strategies to combine human and computer systems. DARPA funded a project beginning in 2006 led by Paul Sajda, a researcher at Columbia University, in New York, to combine the speed of a computer processor and the quickness of the brain's ability to recognize objects. The goal of this project is to accelerate visual searches. Sajda will design a computer system that analyzes the electrical activity of a person's brain while he or she is looking at a series of pictures. The electrical activity of the brain will be recorded from electrodes pasted to the scalp—a common technique known as electroencephalography. When the brain spots something interesting, or recognizes a target it has been searching for, the electrical activity changes, which can be noted by the computer. The system will monitor the human subject while he or she scans a large number of images in a short period of time, and the computer will tag the interesting ones—as revealed by electroencephalography—for more detailed study later. This process saves time by using the speed of the human brain as a "filter" to select the desired images from a large set of candidates.

Combining the talents of brains and computers is an economical solution to the problem of automating visual systems. Computers and computer vision systems have made rapid progress, yet there remain tasks, such as object recognition, that the human brain has mastered but computer algorithms have not. The frontiers of computer vision may not involve replacing human vision as much as partnering with it.

CHRONOLOGY

1948	William Grey Walter (1910–77), a researcher at the Burden Neurological Institute in Bristol, England, devises one of the earliest robots capable of rudimentary sight.
1950s	Dutch company Gatsometer BV develops the earliest traffic enforcement cameras.
1960s	Panoramic Research, Inc., scientists Woodrow W. Bledsoe and his colleagues begin to develop the first facial recognition systems. Peter Hart at Stanford Research Institute and his colleagues also begin working on this technology.
1966–72	Stanford Research Institute scientists, led by Charles Rosen, design and build Shakey, a robot capable of navigating through simple environments.
1975	Eastman Kodak Company engineer Steven Sasson builds the earliest digital camera.
1979	One of the first license plate recognition systems debuts in England.
1993	The U.S. Department of Defense and DARPA assemble a database of images to be used in evaluating facial recognition systems. This project was called FERET, for Facial Recognition Technology.
1997	NASA probe *Pathfinder* reaches Mars and a small rover, Sojourner, capable of autonomous movement, explores the area surrounding the landing site.
2000	The U.S. government sponsors the first Face Recognition Vendor Test to evaluate the available systems.

2004	NASA Mars Exploration Rovers *Spirit* and *Opportunity* land on Mars and began exploring. Computer vision algorithms direct a portion of these journeys.
	DARPA sponsors the first grand challenge. No autonomous vehicle succeeds in completing the 142-mile (227-km) course from Barstow, California to Primm, Nevada.
2005	DARPA sponsors the second grand challenge. The winner and recipient of the $2 million prize is a research team from Stanford University, whose vehicle, *Stanley,* completes a 131.2-mile (210-km) course from Barstow, California, to Primm, Nevada, in six hours, 53 minutes, and 58 seconds.
2007	DARPA sponsors the third grand challenge, known as Urban Challenge. *Boss,* a vehicle designed by researchers at Carnegie Mellon University, wins the race and takes the $2 million prize.

FURTHER RESOURCES

Print and Internet

Enns, James T. *The Thinking Eye, the Seeing Brain: Explorations in Visual Cognition.* New York: W.W. Norton & Company, 2004. As scientists gain a better understanding of the human visual system, the principles they discover may be useful in developing computer vision. This book discusses the biology of human visual processing and cognition.

Hoffman, Donald D. *Visual Intelligence: How We Create What We See.* New York: W.W. Norton & Company, 2000. Everyone's brain displays a remarkable amount of ingenuity in transforming images—regions of brightness and color—into recognizable objects and patterns. This book explores the process of how brains interpret the complex visual world.

Hofman, Yoram, and Hi-Tech Solutions. "License Plate Recognition—A Tutorial." Available online. URL: www.licenseplaterecognition.com. Accessed June 5, 2009. This tutorial offers a simple explanation of how license plate recognition systems work, and presents some of their many applications.

Hubel, David. "Eye, Brain, and Vision." Available online. URL: hubel.med.harvard.edu/index.html. Accessed June 5, 2009. David Hubel is a retired Harvard researcher who shared the 1981 Nobel Prize in physiology or medicine with Torsten N. Wiesel and Roger W. Sperry for their studies of the brain's visual system. Hubel's book, *Eye, Brain, and Vision,* is now available on the Internet, and is one of the best introductions to how vision works.

Jenkins, R., and A. M. Burton. "100% Accuracy in Automatic Face Recognition." *Science* 319 (25 January 2008): 435. This single-page article describes a method to improve recognition accuracy by averaging across a number of photographs.

Johnson, Ryan, and Kevin Bonsor. "How Facial Recognition Systems Work." Available online. URL: computer.howstuffworks.com/facial-recognition.htm. Accessed June 5, 2009. A contribution to the How Stuff Works Web site, this article discusses the current state of face recognition technology and its applications.

Lipkin, Jonathan. *Photography Reborn: Image Making in the Digital Era.* New York: Harry N. Abrams, 2005. Digital techniques have had an enormous impact on photography and imagery. This book describes the rise of digital technology and how it is used in visual art, science, and technology.

Lukáš, Jan, Jessica Fridrich, and Miroslav Goljan. "Digital Camera Identification from Sensor Pattern Noise." *IEEE Transactions on Information Security and Forensics* 1 (2006): 205–214. Available online. URL: www.ws.binghamton.edu/fridrich/Research/double.pdf. Accessed June 5, 2009. The researchers found patterns in digital images that allowed them to trace the digital cameras that took the pictures.

Martínez, Aleix M., and Manli Zhu. "Where Are Linear Feature Extraction Methods Applicable?" *IEEE Transactions on Pattern Analysis and Machine Intelligence* 27 (2005): 1,934–1,944. The researchers developed a test that predicted an algorithm's success or failure in different tasks.

Mataric, Maja J. *The Robotics Primer*. Cambridge, Mass.: MIT Press, 2007. Aimed at a broad audience, including beginning students, this book covers components, motion and navigation, control, and sensors.

Pinto, Nicolas, David D. Cox, and James J. DiCarlo. "Why Is Real-World Visual Object Recognition Hard?" *Public Library of Science Computational Biology* (January 2008). Available online. URL: www.ploscompbiol.org/article/info:doi/10.1371/journal.pcbi.0040027. Accessed June 5, 2009. The researchers review visual recognition methods and criticize overreliance on standard databases.

Retting, Richard A., Susan A. Ferguson, and Charles M. Farmer. "Reducing Red Light Running through Longer Yellow Signal Timing and Red Light Camera Enforcement: Results of a Field Investigation." *Accident Analysis & Prevention* 40 (2008): 327–333. In a study of red-light cameras installed at two major intersections in Philadelphia, Pennsylvania, researchers found that red-light cameras reduced red-light running by more than 90 percent.

Serre, Thomas, Aude Oliva, and Tomaso Poggio. "A Feedforward Architecture Accounts for Rapid Categorization." *Proceedings of the National Academy of Sciences* 104 (10 April 2007): 6,424–6,429. The researchers designed a computer model to simulate early visual processing occurring in the human brain.

SRI International. "SRI International's 'Shakey the Robot' Selected as Robot Hall of Fame Inductee." News release, July 12, 2004. Available online. URL: www.sri.com/news/releases/07-12-04.html. Accessed June 5, 2009. SRI announces that the robot Shakey has been selected for the Robot Hall of Fame.

Web Sites

Defense Advanced Research Projects Agency. Urban Challenge. Available online. URL: www.darpa.mil/GRANDCHALLENGE. Accessed June 5, 2009. Maintained by DARPA, this Web site provides information on the third grand challenge, called Urban Challenge, as well as links to Web sites describing earlier competitions.

Department of Homeland Security. Available online. URL: www.dhs.gov. Accessed June 5, 2009. The Web site of the Department of Homeland Security offers information on travel security and procedures, disaster preparedness and response, defending against terrorist threats, and research on various technologies to enhance the safety and security of the United States.

Face Recognition Vendor Test. Available online. URL: www.frvt.org. Accessed June 5, 2009. This Web site presents the results of the government's evaluations of facial recognition systems. Included are results from 2000, 2002, and 2006.

National Aeronautics and Space Administration. Mars Exploration Rover Mission. Available online. URL: marsrover.nasa.gov/home. Accessed June 5, 2009. This Web site provides an overview of the mission, many interesting facts about the two rovers—*Spirit* and *Opportunity*—the scientific results, and fascinating images from the surface of Mars.

6

Bioinformatics— Using Computers to Explore Biological Data

Computer systems are enormously useful in managing information. As the amount of information grows, the sophistication and efficiency of computer systems become critical factors in the successful application of information technology. Consider a book with 3 billion letters. If it was written in the English language, it would be equivalent to several hundred encyclopedia volumes. A person who could read 10 letters—a couple of words—every second without pausing would need about 10 years to get through the whole book. The book would contain so much information that retrieving any given item would take an extraordinarily long time without the use of an extensive index.

Such a book exists, but it is not written in English, and it does not come with convenient features such as an index. In April 2003, a team of researchers funded by a variety of government agencies completed a project to read and print out the human *genome*—the entire genetic information contained in a person's deoxyribonucleic acid (DNA). Genomes consist of all the information needed for a fertilized cell to grow, divide, and develop into a fully functional adult; in humans, the body is composed of trillions of cells and many specialized tissues and organ systems, such as the heart, liver, and nervous system.

Prior to the completion of the Human Genome Project, biological research in genetics proceeded slowly. Scientists were only familiar with a small part of the genome and had to guess at or spend time looking for what they did not know. Having the genome available has accelerated research because all of the information is available to any researcher. But there is a problem—without a guide or index, looking for a specific bit of information is almost impossible. The human genome consists of 3 billion letters—the sequence of human DNA, written in the nucleotides (also known as bases)—a huge amount of information.

Computer scientists excel in the study of information and the algorithms used to solve problems involved in finding and using information.

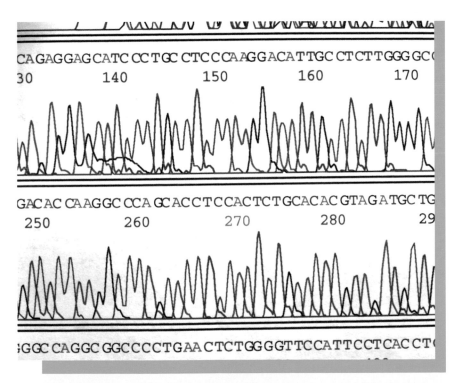

The output of a DNA sequencing machine showing part of the human genome *(SSPL/The Image Works)*

Chapter 4 of this book discussed encrypting and decrypting information, and chapter 5 described the process of automatically extracting features in information-rich images. This chapter examines the techniques of extracting information from an information-rich source—a genome—that is "encrypted" in the "language" of biological organisms. Computers have been used in sciences such as biology for a long time, but the process of managing, searching, and analyzing biological data has a special name—*bioinformatics*. Bioinformatics is an important frontier of science. Although humans are more than just a collection of DNA, genetic information is extremely important in the study of behavior, physiology, and disease.

INTRODUCTION

In 1869, Swiss chemist Friedrich Miescher (1844–95) became the first person to isolate DNA, although no one at the time knew the function or structure of this substance. DNA was termed *nucleic acid* because of its acidic chemical properties and its location, which was mostly in the cell's nucleus (a dense, membrane-bound structure inside cells). Austrian monk Gregor Mendel (1822–84) had studied inherited traits in pea plants, identifying certain heritable factors, later called genes, and had published his work in 1866, but no one paid much attention to it. In 1902, American scientist Walter Sutton (1877–1916) and German researcher Theodor Boveri (1862–1915) suggested that large bodies in the nucleus called chromosomes contain these units of heritance, but it was not until 1944 that scientists realized DNA had anything to do with inheritance and genetics. In that year, Canadian scientists Oswald Avery (1877–1955) and Colin MacLeod (1909–72), and American scientist Maclyn McCarty (1911–2005), reported that genes are made of DNA.

Genes are instructions to make a certain molecule, usually a protein or part of a protein, although a small amount of genetic material is for a type of nucleic acid molecule known as *ribonucleic acid* (RNA). Proteins are the workhorses of a cell, filling a huge variety of jobs a cell needs to stay alive, grow, divide, and perform its function in the body. Many proteins are enzymes, which speed up vital chemical reactions the cell uses to produce energy or make important molecules, and other proteins provide mechanical support or are involved in transporting substances. The type of job a protein does depends on its shape. All proteins consist of molecular units called amino acids that are joined to each other with a strong chemical bond to form a chain. The chain folds into a certain shape that the protein needs to bind reactants and speed up a reaction, or hold and transport specific substances, or strengthen structures within the cell.

Twenty different kinds of amino acid make up proteins. These amino acids have a similar chemical composition, with the same chemical groups attached to a central carbon atom, with one exception—one of the groups is different for each amino acid, distinguishing it from the rest of the amino acids. This attachment can be as simple as a hydrogen atom, as it is in the amino acid glycine, or it can contain a complex structure such as a ring of atoms, as in the amino acids tryptophan, tyrosine, phenylalanine, and histidine. Some of the attachments have electrical charges, whereas others generally do not. These different groups give each amino acid a specific set of properties. The bonded amino acids form chains, and the properties of the amino acids and their interactions govern the shape that a given chain will adopt. For example, many proteins fold into a roughly spherical shape, held together by weak bonds among the amino acids and by their interactions with surrounding water molecules in the solution—some amino acids are attracted to water, and others are repelled. The sequence of amino acids determines the type and location of interactions, which dictates the shape of the protein and therefore its function.

What determines the sequence of amino acids in a protein? This is the job of genes that code for a certain protein or protein segment. DNA stores this information in its composition and structure—the genetic code, which is decoded in order to make the appropriate proteins at the appropriate time in an organism's development. This code is similar to the codes described in chapter 4, in which a message—in this case, the instructions to make a certain protein—are "encrypted" in the gene.

In 1953, James Watson (1928–   ) and Francis Crick (1916–2004) published a short paper in *Nature* on DNA. They discovered the structure of DNA consists of double helix, as shown in the figure on page 15. Each strand of the helix is composed of a chain consisting of sugar molecules known as deoxyribose, along with phosphates and nucleotides, also called bases, which come in four different varieties: adenine (A), guanine (G), cytosine (C), and thymine (T). Holding together the two strands of the helix are weak bonds formed between the bases. Because of their structures, A and T bond together, as do C and G, but no other pairing is possible. This specific base pairing means that a sequence of one strand, say ATTAGC, will bind only to TAATCG, which is known as its complement or complementary strand.

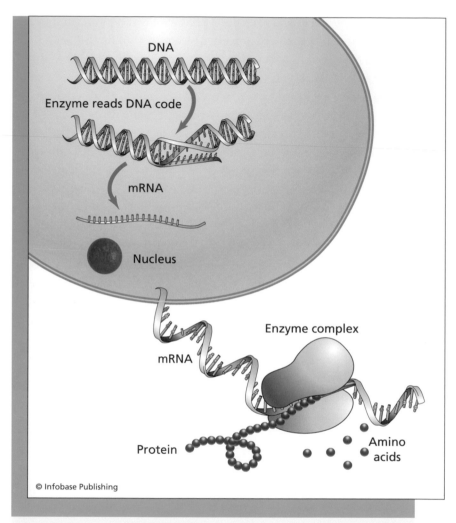

DNA

Enzyme reads DNA code

mRNA

Nucleus

mRNA

Enzyme complex

Protein

Amino acids

© Infobase Publishing

Transcription produces mRNA in the nucleus, which serves as a template for enzymes to make proteins.

The sequence of DNA bases contains the genetic information that specifies proteins and is passed from generation to generation. A double-stranded helix is durable so it lasts a sufficiently long time. Yet it must be copied when the cell divides in two (so that each daughter cell receives a copy), and the genetic code must be read when proteins are to be made. Special enzymes temporarily spread apart the strands so that other enzymes can access the bases. To copy the DNA, enzymes copy each strand, forming two identical double helices.

As shown in the figure, the process of making proteins begins when enzymes assemble a molecule of RNA, called messenger RNA (*mRNA*), whose sequence of nucleotides mirrors the DNA strand that served as a template. (Either strand of a double helix can serve as a template for mRNA.) These mRNA molecules are sometimes called transcripts because enzymes "transcribe" DNA to make mRNA. Molecules of mRNA exit the nucleus, attach to gigantic complexes of other RNA and proteins that "read" or "translate" the code and assemble the appropriate protein.

In the 1960s, Crick, Sydney Brenner (1929–), François Jacob (1920–), Marshall Nirenberg (1927–), and other researchers discovered how to read the genetic code. Three consecutive bases in a gene form a *codon,* which codes for specific amino acids. For example, GGA codes for glycine, and CAC codes for histidine. There are 64 possible codons, each of which codes for one of the 20 amino acids, or a special instruction called stop, which terminates the protein-making process. (Most amino acids have more than one codon assigned to them.) The sequence of DNA gets turned into a sequence of amino acids—a protein with a specific function. Special molecules or enzymes control and regulate this process, determining which protein gets made at what time.

All individuals, except identical twins, contain a unique set of genes and DNA. Most cells in the human body contain 23 pairs of chromosomes in the nucleus, plus a small amount of DNA in structures called mitochondria. The total amount of DNA is about 3 billion bases. (Genes vary in size, but average about 3,000 bases.) One chromosome of each pair is inherited from the mother and one from the father, so that offspring contain a mixture of genes from both parents. Genes come in slightly different versions called *alleles,* which have slightly different sequences and therefore code for slightly different proteins. This variability, along with gene shuffling that occurs during reproduction, accounts for much (though not all) of the variability in human behavior, appearance, and physiology. As in all organisms, the process of evolution selects the best set of alleles for a given environment or conditions because successful organisms will survive and reproduce, passing along their alleles to their offspring.

Forensics experts can sometimes solve crimes by identifying suspects based on the unique DNA sequence of individuals. (The only people with the same DNA are genetic twins, who arise from the same fertilized egg cell.) Biologists study genetic information because genes determine or influence so many biological functions and properties, including evolution, traits such as height and body type, the structure and function of organs and systems, susceptibility to disease, and response to medications.

Throughout the 1970s and 1980s, scientists studied genes and specific alleles that governed specific traits, but these genes were enormously difficult to find. Another problem biologists faced was the sheer magnitude of the 3 billion DNA bases, most of which had not been sequenced at the time. The importance of genetics in biology and the treatment of disease motivated an ambitious project, the Human Genome Project, to sequence all human DNA. In 1990, two United States government agencies—the National Institutes of Health and the Department of Energy—joined international research teams in the quest to sequence all 3 billion bases. The project was finished on April 2003, at a cost of about $2.7 billion. Francis Collins, director of the National Human Genome Research Institute, remarked, "The availability of the 3 billion letters of the human instruction book could be said to mark the starting point of the genomic era in biology and medicine."

The completion of the Human Genome Project was an important event, but it is only one step, albeit a large one, in genetics research. The project gave researchers the full sequence of one specific human genome. (DNA from only a few anonymous individuals was used in the project, and scientists patched the different contributions into one full sequence. The Human Genome Project sequence is not any individual's sequence, but instead comes from bits and pieces of a few individuals.) This sequence is simply a long list of bases; it says little about variability in humans, does not specify which sequences are used for what gene, and does not in general describe the function of any DNA segment at all. The job of making sense of this enormous amount of information belongs to biologists, who at once realized they needed help from computer scientists having expertise in managing, storing, and searching data.

SEQUENCES, GENOMES, AND PROTEINS

Even before the Human Genome Project's completion, biologists knew that much of the human genome did not contain genes. Only about 2

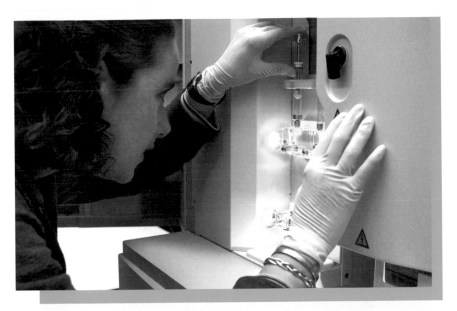

Researcher checking a DNA sequencing machine *(Maggie Bartlett/ NHGRI)*

to 3 percent of human DNA sequences are genes, and the remainder is sometimes described as "junk DNA" or noncoding DNA (meaning DNA that does not code for proteins or RNA). Some of this DNA consists of sequences repeated many times (why these repetitive sequences exist is not well understood), but the term *junk DNA* is not accurate because researchers have found important sequences in these regions. Some of these sequences regulate gene expression—the process of transcribing genes into mRNA and assembling the proteins. Because researchers are just beginning to understand the genome, the Human Genome Project needed to sequence all the bases.

The genomes of many other organisms have also been sequenced. Genomes of animals such as the cow, rabbit, rat, mouse, dog, cat, and chimpanzee, along with many bacteria and other simple organisms, have been completed or are near completion. Biologists are interested in studying the genetics of these organisms for many reasons, including the possibility of learning something about humans by comparing human genes with those of other organisms, as described in the following section.

Data must be organized in order to be useful. A large collection of data is usually stored in a computer database, which is a set of records

National Center for Biotechnology Information

Long before the Human Genome Project and other genome sequencing projects began, biological and biomedical scientists had to deal with huge quantities of data. Biology is a data-rich science—biologists make a large number of observations and conduct many experiments that produce a lot of data, from which generalizations gradually emerge (for the lucky and diligent researcher). The U.S. National Library of Medicine (NLM), established in 1836 as the Library of the Office of the Surgeon General of the Army, stores and indexes journals and research materials. After the development of computers, NLM staff became proficient with computer databases. Facing a deluge of data from genetics research and associated biotechnology, the U.S. government established the National Center for Biotechnology Information (NCBI) on November 4, 1988, as part of the NLM.

More than 1,500 genomes from different organisms can be accessed at the NCBI Web site. The list includes genomes of the human, the mouse, the dog, the chimpanzee, the mosquito, the fruit fly, and more than 600 species of bacteria. NCBI also distributes information from a number of other important databases, including the Online Mendelian Inheritance in Man, a database of genes and genetic disorders, the Cancer Genome Anatomy Project, which contains information about genes associated with cancer, and the Molecular Modeling Database of three-dimensional protein structures.

In addition to maintaining and supporting databases, the NCBI educates and trains scientists on the effective use of these tools in science. Researchers at the NCBI develop computer algorithms to make searches more efficient and productive, and engage in their own investigations and analyses. As the data continue to pour in, services provided by the NCBI have become increasingly important in biology and bioinformatics.

stored and organized in a certain manner. The organization allows quick access to the data for easy searching, and provides links to related records. For instance, each state's department of motor vehicles maintains a database of registered vehicles, which allows authorities to find the owner of a car with a certain license plate number.

Genome sequences and data are housed in specially designed databases. Computer programs allow scientists to browse or search the data, and in some cases, annotate the records by contributing information. For example, a researcher may discover that a region or sequence contains a gene that performs a certain function, so this information can be added to the database record.

The Internet is an invaluable tool for bioinformatics. In an article, "Bioinformatics in the Post-Sequence Era," published in a 2003 issue of *Nature Genetics,* researchers Minoru Kanehisa and Peer Bork wrote that the Internet "has transformed databases and access to data, publications and other aspects of information infrastructure. The Internet has become so commonplace that it is hard to imagine that we were living in a world without it only 10 years ago. The rise of bioinformatics has been largely due to the diverse range of large-scale data that require sophisticated methods for handling and analysis, but it is also due to the Internet, which has made both user access and software development far easier than before."

Web sites allow researchers from all over the world to access the databases as part of their research projects. Several Web sites house databases containing the human genome sequence; for example, the University of California at Santa Cruz hosts a Web site (genome.ucsc.edu) containing the sequence of the human genome and other genomes. One of the most important managers of genome data is the National Center for Biotechnology Information, as described in the sidebar on page 164.

Since the Human Genome Project contains data from only a small number of individuals, the sequence is not of much use in studying genetic variability. But the sequence gives researchers a look at every human gene (at least one allele of each gene) for the first time. The only problem is that the genes make up only 2 to 3 percent of the total, and are buried within 3 billion bases of human DNA.

Most estimates for the number of genes in the human genome were around 100,000 before the completion of the Human Genome Project. (An informal betting pool arranged at a scientific meeting held in 2000 at Cold Spring Harbor Laboratory, in New York, received bets ranging from about 26,000 to 150,000.) Researchers had already found some genes using difficult and time-consuming laboratory and genetic methods, developed prior to the Human Genome Project. For example, Nancy Wexler and a team of geneticists spent more than a decade tracking down the gene that causes Huntington's disease, a progressive brain disorder, before finally succeeding in 1993.

But even after the completion of the Human Genome Project, researchers were not sure of the identity of every human gene. Scientists had the sequence of the entire human genome in their computers, but nature does not put a flag or red marker next to the genes. Researchers have had to find the genes themselves. This task is an important part of bioinformatics.

Finding genes in simple organisms such as bacteria are easier than in the human genome, because bacterial genes often begin near specific sequences that are involved in regulating expression. In more complex organisms, the regulatory sequences and other associated gene markers are not always so close. Another complication is that genes contain introns—noncoding segments that are snipped out of the mRNA. To recognize a gene, bioinformatics researchers often analyze a region and study the possible proteins for which this sequence would code. The amino acid sequence of the protein would depend on exactly where the gene started, so a variety of possible beginning points is identified, and then the codons are read until the next stop codon.

Additional information comes from protein databases. Scientists already know the sequence of a large number of proteins, and many of these proteins have characteristic sequences. For example, a leucine zipper is a pattern found in certain proteins in which the amino acid leucine is found at every seventh position. Researchers can also use mRNA sequences, if known, to help identify genes.

After the completion of the Human Genome Project, estimates of the number of genes dropped considerably, with gene-finding programs initially predicting less than 30,000. (The winner of the betting pool was announced in 2003—Lee Rowen of the Institute of Systems Biology in Washington won with his bet, 25,947, which was the lowest prediction.)

Although researchers still do not know the exact number of human genes, bioinformatics algorithms continue to be refined. Some of the most recent algorithms gain an advantage from the similarity of DNA sequences among related species. For example, chimpanzee and human DNA sequences are about 96 percent identical.

A recent approach that compares genomes and proteins of related species has led to an additional downward revision of the number of genes in the human genome. Michele Clamp and Eric Lander of the Massachusetts Institute of Technology, along with their colleagues, scrutinized all of the genes previously identified in the human genome. Some of these regions contained sequences that appear to be capable of producing a viable protein, say protein X, but Clamp and her colleagues checked all of the databases for a protein similar to X. Critical genes that help an organism survive and reproduce are conserved, even as organisms evolve, accounting for the similarity in genetic sequences among related animals such as chimpanzees and humans. When Clamp and the other researchers failed to find a similar protein, then this putative gene must either be unique to humans or it is not a coding region after all. But since the chimpanzee and human genomes are so alike, they mostly contain the same genes—subtle differences exist in the genes of the two organisms, but there are not that many wholly different genes. The researchers eliminated several thousand genes with this process, bringing the total to 20,500. This work, "Distinguishing Protein-Coding and Noncoding Genes in the Human Genome," was published in 2007 in the *Proceedings of the National Academy of Sciences.*

The exact gene count can only be verified with biological experiments and measurements to prove which sequences are genes that are actually expressed. But bioinformatics has made an important contribution in pinpointing the most likely candidates to study.

LEARNING FROM MICE AND CHIMPANZEES—COMPARING GENOMES AND PROTEINS

The research of Clamp, Lander, and their colleagues indicate the usefulness of comparative genomics—studies that analyze and compare genomes of different organisms. Comparative genomics is possible because the genomes of many organisms have been sequenced.

Scientists have been using animals in biomedical studies ever since Darwin's ideas about evolution, published in 1859, indicated a

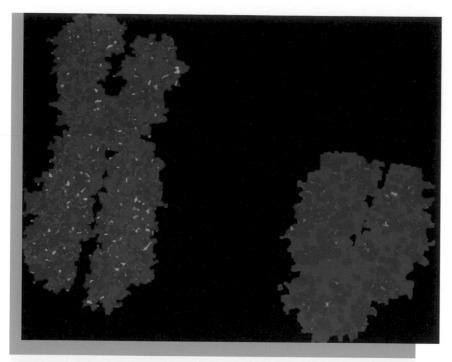

Electron microscope image of human X and Y chromosomes *(Dr. K. G. Murti/Visuals Unlimited)*

close relationship between animal and human physiology. Evolution tends to preserve successful genes as well as physiological systems. For example, the physiology of all mammalian species is similar because these properties were inherited from a common ancestor, and modified by subsequent evolution. The modifications can be subtle or, in some cases, quite drastic, though animals that descended from a common ancestor a relatively short time ago (which in evolution is millions of years) are usually much alike. Crucial systems tend to be especially preserved; for instance, the genetic code is the same for virtually all organisms.

In genetics, the mouse is a popular choice for study. Mice reproduce quickly, amassing many generations in a short period of time, which is ideal for the study of genetic transmission, and mice having specific characteristics can be produced. Being mammals, mice have similar physiological systems to humans, despite the physical differences. Even the few physiological differences can be beneficial—for reasons that are not well understood, mouse genes are easier for experimenters to manipulate. Researchers can even produce a mouse lacking one or more specific genes; such mice are known as "knock-outs," since the gene has been knocked out of the genome. Knock-out experiments let researchers study the function of genes, since animals lacking a gene will lack that particular function.

In 1999, an international team of researchers began sequencing the mouse genome. Although this project began much later than the Human Genome Project, it finished quickly, in 2002, mostly because of improvements in sequencing technology rather than a smaller genome size. Mice have 20 pairs of chromosomes and 2.5 billion bases of DNA, only slightly smaller than the human genome's 23 pairs of chromosomes and 3 billion bases.

Bioinformatics tools such as search algorithms can explore, or "mine," these data. These searches reveal possible genes and other important sequences, and compare organisms for similarities and differences, providing clues about how evolution works as well as helping to understand the physiological systems of different organisms. An important bioinformatics tool called Basic Local Alignment Search Tool (BLAST), described in the sidebar on page 170, is used to query a database of DNA or protein sequences. For example, if a gene is discovered in a mouse, researchers will want to search the human genome to see if there is a gene with a similar sequence.

The Mouse Genome Sequencing Consortium, an international research team, reported in an article titled "Initial Sequencing and Comparative Analysis of the Mouse Genome," published in a 2002 issue of *Nature,* "Approximately 99% of mouse genes have a homologue in the human genome." A homologue indicates similarity. This finding puts added emphasis on experiments in mouse genetics, since for almost every gene in the mouse there is a similar gene in humans that might have a related function. Researchers can learn a lot about human genes by doing experiments in mice that would not be possible or ethical in humans.

For example, Ronald A. DePinho, Lynda Chin, and their colleagues at Dana-Farber Cancer Institute, in Boston, Massachusetts, recently developed a more effective mouse "model" of cancer. Scientists often study complex problems by developing simpler models that incorporate the essential features of the subject under study. A model may be a small-scale version of a structure, or it may be a computer program that mimics the behavior of some object or structure. Biologists often use animal models, which are laboratory animals such as mice that get

Searching Databases Is a BLAST

Even before the Human Genome Project began to accumulate large amounts of data, the sequences of some proteins and large segments of DNA from a variety of organisms had been acquired. To search and compare these sequences, NIH researchers Stephen Altschul, Warren Gish, Webb Miller, Eugene Myers, and David J. Lipman published an algorithm in an important paper, "Basic Local Alignment Search Tool," in a 1990 issue of the *Journal of Molecular Biology*. One measure of the importance of a science paper is its number of citations; when scientists report their results in a published paper, they cite references—previously published papers—that influenced their work or help to explain it. An influential paper will receive a lot of citations. Scientists cited the BLAST paper published by Altschul and his colleagues about 10,000 times in the 1990s, making this paper the most highly cited science paper of that decade, according to *Science Watch*, a scientific newsletter published by Thomson Scientific.

Databases such as those maintained by the NCBI have BLAST software installed on their computers. BLAST searches begin when a researcher enters a sequence, usually representing a protein or a gene, and specifies the appropriate database to search. The database may as large as the 3-billion-base human genome, as would occur when a researcher wishes to search for a human gene that is similar to a mouse gene he or she has just discovered. BLAST may find

diseases similar to humans. By studying the animal model, and discovering the most effective treatment strategy, researchers hope to apply this knowledge to the medical treatment of human patients.

The Dana-Farber Cancer Institute researchers developed gene knock-out mice that lacked genes important for stabilizing DNA mola perfect match, but this is unlikely—mice and human genes generally have a number of differences in the sequence, even if they perform the same function. More likely, the search will turn up sequences that are similar but not identical.

An important aspect of the algorithm is to measure, or score, similarity. A lot of sequences in the database may have stretches that are more or less similar to the given sequence, but BLAST tries to find the best one based on its scoring measures. The program first compares two sequences, looking for a short stretch where they are identical. After aligning the two sequences based on this region, the algorithm looks for more similarities in the surrounding region, accounting for the possibility that further sequence matches may require realignment. For instance, two DNA sequences that match in a region such as ACCGAT will be aligned at that spot. Additional similarity around this region will increase the similarity score.

Measuring similarity may not seem difficult, but with a database containing billions of entries, the task demands an efficient algorithm such as BLAST. When a researcher runs the program, he or she will obtain a list of matches, ranked by their score. The algorithm may find one sequence that has a much higher score than the rest, which would suggest that the researcher has found a good match, or it may not, indicating a lack of any good matches. But similarity between two genes or gene segments from different organisms does not necessarily mean that these two sequences have the same function. Establishing functional similarity may require further testing and experiments.

ecules. Although the double helix structure of DNA tends to be stable, chemicals and stimulants such as ultraviolet light can destabilize DNA, leading to genetic mutations. Special proteins in the cell protect DNA from these attacks and quickly repair any damage. In mice lacking the genes that code for some of these proteins, the animals developed severe cases of certain types of cancer. Cancer, the second leading cause of death in the United States after heart disease, is a disease in which cells grow and divide uncontrollably, often invading other tissue and forming a mass called a tumor. Richard S. Maser, DePinho, Chin, and their colleagues examined the DNA of the mice knock-outs and discovered genome instabilities with a striking similarity to those in cells of certain human tumors. These instabilities included large rearrangements of chromosomes. The researchers published their findings, "Chromosomally Unstable Mouse Tumours Have Genetic Alterations Similar to Diverse Human Cancers," in *Nature* in 2007. Further study of these mice may help scientists find important clues about how cancer arises.

In terms of evolution, chimpanzees are much closer to humans than mice are—scientists believe humans and chimpanzees evolved from a common ancestor about 5 million years ago, whereas mice and humans probably diverged more than 75 million years ago. Researchers cannot perform many experiments on chimpanzees because of their size, rarity, and the belief of many people that these animals are too similar to humans. But the chimpanzee genome can be studied, and the Chimpanzee Sequencing and Analysis Consortium, an international research team, finished sequencing the chimpanzee genome in 2005. Chimpanzees have 24 pairs of chromosomes and about 3 billion bases of DNA.

By studying the genetic differences between chimpanzee and humans, biologists will gain a better understanding of what causes the differences in physical characteristics between the two species. Other important questions to answer concern genes involved in nervous system development and function. The chimpanzee brain is quite similar to humans in anatomy and structure, yet humans possess the intellectual capacity for technology, science, and language that is lacking in these otherwise intelligent animals.

A simple sequence comparison between chimpanzee and human genomes shows they are about 96 percent identical, as mentioned earlier. But a recent study comparing the human and chimpanzee genomes took a broader perspective. University of California at Santa Cruz researchers Katherine S. Pollard, David Haussler, and their colleagues searched for sequences that are highly conserved in vertebrates—most animals with backbones have these sequences in their DNA, which means that this DNA probably has a critical function because it has not changed over millions of years of evolution. Yet the researchers discovered 202 of these conserved sequences show changes in human DNA but not in chimpanzee DNA, so these sequences have undergone evolution in humans at a higher rate than in chimpanzees. Many of these sequences are noncoding. This finding suggests that a lot of the genetic distinction between humans and chimpanzees does not lie in the coding part of genes—and thus the proteins they encode—but instead lies in their regulation, or, in other words, which proteins are made, and when. This research, "Forces Shaping the Fastest Evolving Regions in the Human Genome," was published in 2006 in *Public Library of Science Genetics.*

NETWORKS OF GENES AND DNA SEQUENCES

As bioinformatics and genetics research expands, researchers must tackle increasingly difficult and complex issues. BLAST and other algorithms perform valuable services such as searching the data and finding the genes, but this research does not generally reveal the functions of these genes, nor does it explain the function of noncoding DNA.

The importance of noncoding DNA is becoming increasingly evident, as indicated in research such as that by Pollard, Haussler, and their colleagues. To broaden the scope of DNA investigations, a team of researchers consisting of 35 groups from all over the world pooled their resources and formed the Encyclopedia of DNA Elements (ENCODE) consortium. The National Human Genome Research Institute, a branch of the NIH, organized this consortium to carry out a meticulous examination of all functional elements in a small section of the human genome, amounting to about 1 percent of the total. A functional element is any DNA having a specific function, such as a gene that codes for a protein or RNA molecule, a sequence that regulates gene expression, or a sequence that maintains the structure and integrity of chromosomes. One percent of the genome is not a lot in terms of percentage, but it consists of about 30 million bases and was considered a feasible goal for the initial four-year project, launched in 2003.

The ENCODE project combined laboratory experimental techniques with computational analyses. Some of the experimental techniques include identifying when certain genes are expressed by measuring either the protein product or the mRNA transcripts. But with thousands of genes, measuring all of these products or transcripts at the same time is exceptionally difficult. Researchers often use a tool such as a microarray, which is an array consisting of hundreds or even thousands of tiny wells or containers. Each well contains molecules that bind a certain kind of protein or mRNA. For proteins, the binding molecules may be antibodies, which are molecules of the immune system that latch onto any kind of invading molecule; researchers can produce antibodies to certain human proteins by, say, injecting a human protein into an animal, and then collecting the antibodies that animal's immune system makes against the "invader." In the case of mRNA, researchers can produce an RNA molecule that is the complement of a specific mRNA, and when the two meet they will tend to bind together. To use the microarray, a researcher pours a cell or tissue sample over the wells, each of which isolates and measures the amount of a specific protein or mRNA transcript.

Microarrays offer a fast way of isolating a large number of proteins or mRNA transcripts at the same time, and a variety of other laboratory measurement techniques, or assays, identify important sequences such as regulatory elements and other active sites. In addition to gathering this data, the ENCODE project developed bioinformatics algorithms to study the targeted regions of the human genome in more depth. An important component of ENCODE's bioinformatics research was the analysis of comparable regions in the genomes of other organisms. As discussed above, such comparisons often allow researchers to get a better understanding of the function of a given gene or DNA element, as well as identifying conserved sequences. ENCODE project researchers used sequences from 14 mammalian species and an additional 14 vertebrate species.

In 2007, ENCODE published its findings. The results were surprising. Instead of broad swaths of "junk" DNA interspersed with a few genes, the researchers discovered the majority of DNA may be transcribed into RNA. No one is yet certain what function these RNA molecules serve, but this activity suggests that little of the human genome may actually be unused. These sequences seem to act together in a vast network of interacting elements. Some of these sequences have not been conserved during evolution, which suggests more variability among species than had been originally believed. Studying this complex network will require much research, including bioinformatics. The ENCODE Project Consortium published its results as "Identification and Analysis of Functional Elements in 1% of the Human Genome by the ENCODE Pilot Project" in a 2007 issue of *Nature*. Researchers reported additional findings in 28 papers published in the June 2007 issue of *Genome Research.*

GENETIC VARIATION

Studies of the genome encompass all the functional elements in DNA, but everyone's DNA is different (except identical twins). Although most individuals contain all or virtually all of the functional elements, the sequence of these elements has slight variations. These individual differences are important determinants in physical characteristics, such as eye color, height, and the tendency to contract certain diseases such as cancer, as well as the tendency to respond to certain types of treatment.

Genetic variation may come in different forms. DNA from different individuals is much alike, perhaps as much as 99 percent identical, although researchers presently lack a large number of fully sequenced individuals to compare, so no one is sure of the exact figure. But when comparing DNA sequences of small regions from the same location—say, at a certain place on chromosome 2 or 5—different individuals have the same sequence, except that every once in a while a single base, or nucleotide, is different. Perhaps a C is present on one individual, whereas a T is present on another. This occurs at about one nucleotide in 100. The site of this kind of variability is known as a *single nucleotide polymorphism* (SNP) because this nucleotide has a variable form (polymorphism) in different individuals. (SNP is usually pronounced "snip.") An SNP, illustrated in the figure on page 176, may occur in genes or in noncoding regions.

SNPs are not the only places in the human genome where individual DNA varies. A lot of human DNA contains repeated sequences; although the function, if any, of much of this repetitive DNA is unknown, researchers do know that the number of repetitions varies among individuals.

Research on genetic differences has concentrated on SNPs, but scientists such as Jan O. Korbel and Michael Snyder of Yale University, in New Haven, Connecticut, have uncovered evidence of variations with a size larger than single nucleotides. Snyder and his colleagues compared the sequences of two females, one of African descent and one of European descent. The researchers cut the DNA into small pieces about 3,000 nucleotides long, and then scanned the pieces for differences by

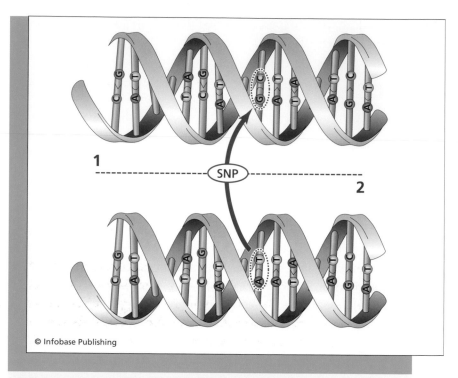

© Infobase Publishing

SNPs consist of a difference in a single base of DNA.

using special, rapid sequencing techniques. In addition to SNPs, the researchers found areas where sequences have different arrangements and structures. This research, "Paired-End Mapping Reveals Extensive Structural Variation in the Human Genome," was published in a 2007 issue of *Science*.

Bioinformatics researchers maintain large databases containing these variations and where they have been found. For example, the NCBI's database, dbSNP, houses more than 12 million sequence variations, including SNPs and other sites of individual variability. Bioinformatics algorithms will continue to be important in identifying and studying these differences.

GENES THAT CAUSE CANCER

Cancer is a disease in which genes and genetics play a strong role. Young, growing individuals need many new cells, and adults need to replace damaged or dead cells, but cell growth and division must stop once the desired number of cells has been reached. Certain genes code for proteins that monitor and carefully regulate the process. Sometimes, however, these genes become damaged, as occurs when components of cigarette smoke or a strong dose of ultraviolet radiation causes mutations that alter the DNA sequence. When this happens, the protein coded by these genes no longer functions because it is made with the wrong sequence.

In some cases, people inherit defective genes from their parents, and these faulty genes, along with mutations acquired from smoking or sun damage, destroy the regulatory process in certain cells. These cells begin to divide uncontrollably, and cancer occurs if the cells spread out and invade the rest of the body. The type of cancer depends on where it begins and which genes are involved.

Researchers are seeking to identify these genes in order to study them and to discover the most effective treatments for each type of cancer. P. Andrew Futreal and his colleagues at the Wellcome Trust Sanger Institute in Hinxton, England, cataloged these genes in 2004, listing 291—about 1 percent of the total number of human genes—that are involved in some form of cancer. This paper, "A Census of Human Cancer Genes," was published in a 2004 issue of *Nature Reviews Cancer.*

Bioinformatics contributes to this research by searching the genome to find the location of these genes and the sequences that regulate them. This effort has been aided in 2006 when the NIH launched the Cancer Genome Atlas project. The goal of this project is to find all cancer-associated genes and sequences by scanning all 3 billion bases of human DNA. To accomplish this ambitious goal, researchers must employ their bioinformatics algorithms on DNA obtained from cancer patients. By comparing and analyzing the DNA sequences of cancer patients with one another, as well as with the DNA of people who do not have cancer, researchers will be able to pinpoint the essential genetic features associated with this disease.

Other prominent diseases such as heart disease also involve a combination of environmental factors (such as diet and exercise) and genes, some of which account for a greater or lesser tendency to contract the disease. Genetic components of any disease may be dissected with bioinformatics along with medical observation and biological experimentation.

But even more ambitious projects await bioinformatics. Computer analysis of genomes opens the pathway to study an organism in its entirety.

A DIGITAL REPRESENTATION

The main thrust of bioinformatics is to manage the huge quantity of data generated by sequencing projects and to mine these data in order to extract as much meaningful information as possible. Computers are essential elements in creating the databases and running the algorithms needed to accomplish these tasks. But computers can do much more. Not only are computers involved in search and analysis operations, but computer programs can also use the information generated by these operations to simulate complex biological processes such as genetics and evolution.

Chapter 5 of this volume described how digital representations changed photography and the field of computer vision. A digital representation of a picture or image is a series of numbers representing the brightness and color of dots, or pixels, composing the image. These numbers contain the data of the image in digital format, which can be stored in a computer file and transformed, as needed, back to a visual image by a program that reads these numbers and "draws" the picture.

Databases represent the human genome as a long series of bases, or letters. These data contain a great deal of biological information regarding the growth, development, and evolution of organisms. Scientists can make excellent use of these representations to simulate, or mimic, organisms with the aid of special computer programs. This process is similar to artificial neural networks—implementing the principles of the brain and its neural networks in computer programs creates useful algorithms, but is also a productive technique for studying how the brain works, if the implementation is realistic.

Simulations of genetics include such applications as "artificial life." Artificial life is the study and development of computer programs and other techniques to mimic life forms, incorporating details of physiology such as metabolism (the production of energy) and genetics. Suppose, for example, that a researcher writes a complex computer program to create a kind of "world" inhabited by creatures. These creatures may need certain resources such as food and water to survive, and pass along inherited traits to their offspring by rules similar to genetics. If the simulation incorporates much of what is known about physiology and genetics, even if in a simplified form, a researcher can study processes such as evolution while observing the artificial creatures evolve and adapt to various situations. Such programs can also be entertaining,

Artist's conception of a digital representation of a human being *(Mike Agliolo/Photo Researchers, Inc.)*

and are the basis for video games such as Creatures, developed in the 1990s by British computer scientist Stephen Grand and his colleagues. Scientific research can be a lot of fun for many reasons!

Other computer applications involve detailed models of an existing organism either in part or in whole. Even small organisms contain many genes and cells, making it infeasible to model the whole organism or even a large component in any detail, but models incorporating genetic and DNA information can include a small number of realistic cells. Such models allow researchers to test hypotheses and hunches without having to do difficult or laborious experiments on "real" systems. Jasmin Fisher of the Swiss Federal Institute of Technology in Lausanne, Switzerland, and her colleagues recently created a model of cells of *Caenorhabditis elegans,* which is small worm about 0.04 inches (0.1 cm) in length. The model is dynamic—it follows the progression of development as certain cells mature.

Many cells in a young, developing organism start out in a general state and gradually assume special characteristics, becoming, for instance, skin cells, heart cells, or brain cells, each of which has a different function and unique properties. Underlying this progression from general to specific form is the exchange of signals that results in the activation of certain genes. *Caenorhabditis elegans* has been widely studied—it is an important model system in biology—and Fisher and her colleagues based their model on the genetics and physiology of this worm and its vulval system (involved in reproduction).

As described in their report, "Predictive Modeling of Signaling Crosstalk during *C. elegans* Vulval Development," published in 2007 in *Public Library of Science Computational Biology,* the researchers showed which genes and gene interactions are important if the cells are to develop correctly. Fisher and her colleagues also confirmed the findings of their model with experiments. The researchers wrote, "Analysis of our model provides new insights into the temporal aspects of the cell fate patterning process and predicts new modes of interaction between the signaling pathways involved. These biological insights, which were also validated experimentally, further substantiate the usefulness of dynamic computational models to investigate complex biological behaviors." These processes are interesting to study for the biological knowledge they reveal, but this research is also worth pursuing because similar events occur during the transformation of normal cells to cancerous cells.

As bioinformatics continues to explore genomes and protein sequences, more physiological and genetic details will emerge. This knowledge is extremely useful in computational biology, where the study of models and simulations give researchers the opportunity to test hypotheses and explore biological systems *in silico*—with computer technology.

CONCLUSION

Algorithms such as BLAST, which search the huge and growing collection of sequenced genomes, demonstrate the great utility of computers in biology. Without the aid of computers, even the mere list of 3 billion letters—the sequence of all human DNA bases—would be bulky and difficult to manage. Finding genes would take decades.

But now that bioinformatics is getting a handle on the location and function of genes, a new set of challenges has arisen. Individual variation is a critical aspect for evolution and the distribution of traits in a population, including genetic diseases and the tendency to contract disorders such as cancer and heart disease. But a small number of samples—the genomes of just a few members of the species—is not sufficient to conduct reliable studies of variation. Some alleles are rare, found in only a small percent of the population, yet are quite important. For example, a genetic trait occurs in about 6 percent of people that renders them insensitive to certain types of painkillers known as opiates; for these people, opiate painkillers offer no relief.

Amassing and studying a large number of samples is burdensome when the genome is large, as it is in humans. But the benefits of understanding individual variation at the genetic level would be enormous. In January 2008, an international research consortium announced the 1000 Genomes Project, an ambitious project to sequence the genomes of at least 1,000 people. Sponsors of the project include the NIH, Wellcome Trust Sanger Institute, in Hinxton, England, and the Beijing Genomics Institute, in China. Project managers expect three or more years will be required to finish.

To study the full breadth of human variation, the project's 1,000 participants must come from a broad range of backgrounds and ethnicities. These volunteers will be anonymous, as were the donors in the Human Genome Project, to protect their privacy. The 1000 Genomes Project aims to uncover most of the alleles and variants in human genetics, including those that are so rare that they are found in only one person out of 100. Variations, whether common or rare, will be interesting to study in terms of distribution (who has which variant) and the possible associations with certain traits, behaviors, and susceptibility to disease.

At 3 billion bases per participant, the 1000 Genomes Project can expect to handle 3 trillion bases of data in all. This amount of data is more than researchers have stored in DNA databases in the previous two decades. Computers and bioinformatics will become even more important, requiring faster, more effective algorithms to search and mine this enormous quantity of data. There is a growing need for a new kind of highly versatile researcher—one who is skilled in both biological science and computer science.

Genetic differences and individual variation are important to study and understand, but they do not define the whole person. Humans are not simply a collection of DNA sequences—life is richer than that. But there remains a wealth of scientific knowledge in genetics, evolution, cancer, and related subjects that is yet to be discovered. Bioinformatics is in its early stages, but scientists already have the entire human genome stored away in computer databases, and there is still much information left to explore.

CHRONOLOGY

1866	Gregor Mendel (1822–84), an Augustinian monk, publishes his experiments with pea plants and discusses how traits are inherited, though his work is mostly ignored.
1869	Swiss chemist Friedrich Miescher (1844–95) becomes the first person to isolate and analyze DNA.
1900	German genetist Carl Correns (1864–1933), Dutch botanist Hugo de Vries (1848–1935), and Austian scientist Erich von Tschermak-Seysenegg (1871–1962) (whose grandfather had once taught Gregor Mendel), rediscover Mendel's work.
1902	American scientist Walter Sutton (1877–1916) and German zoologist Theodor Boveri (1862–1915) propose that chromosomes contain the units of inheritance.
1909	Danish botanist Wilhelm Johannsen (1857–1927) coins the term *gene*.
1944	Canadian scientists Oswald Avery (1877–1955) and Colin MacLeod (1909–72), and American scientist Maclyn McCarty (1911–2005), show that genes are made of DNA.
1953	American biologist James Watson (1928–) and British scientist Francis Crick (1916–2004) propose that the structure of DNA is a double helix. They base their findings in part on experiments performed by British scientists Maurice Wilkins (1916–2004) and Rosalind Franklin (1920–58).

1960s	Crick, South African biologist Sydney Brenner (1929–), French biologist François Jacob (1920–), American scientist Marshall Nirenberg (1927–), and their colleagues decipher the genetic code.
1976	Belgian biologist Walter Fiers and his colleagues are the first to sequence the complete genome of an organism, in this case a virus, MS2.
1988	The United States establishes the National Center for Biotechnology Information to develop and maintain databases and to engage in bioinformatics research.
1990	Two U.S. government agencies, the National Institutes of Health (NIH) and the Department of Energy, along with teams of international researchers, begin the Human Genome Project.
	NIH researchers Stephen Altschul, Warren Gish, Webb Miller, Eugene Myers, and David J. Lipman publish the BLAST algorithm in a paper, "Basic Local Alignment Search Tool," in a 1990 issue of the *Journal of Molecular Biology*.
2002	The Mouse Genome Sequencing Consortium, an international research group, finishes sequencing the mouse genome.
2003	Human Genome Project researchers finish sequencing the human genome.
2005	The Chimpanzee Sequencing and Analysis Consortium, an international research team, finishes sequencing the chimpanzee genome.
2007	The Encyclopedia of DNA Elements (ENCODE) consortium, a team of researchers consisting of 35 groups from all over the world, finish conducting a painstaking examination of all

functional elements in about 1 percent of the human genome.

2008 An international research consortium including the NIH, Wellcome Trust Sanger Institute, in Hinxton, England, and the Beijing Genomics Institute, in China, announces the 1000 Genomes Project. The goal is to sequence the genomes of 1,000 people.

FURTHER RESOURCES
Print and Internet

Altschul, Stephen F., Warren Gish, Webb Miller, Eugene W. Myers, and David J. Lipman. "Basic Local Alignment Search Tool." *Journal of Molecular Biology* 215 (1990): 403–410. Available online. URL: www.cs.umd.edu/~wu/paper/blast_ref.pdf. Accessed June 5, 2009. The researchers describe the BLAST algorithm.

Clamp, Michele, Ben Fry, Mike Kamal, Xiaohui Xie, James Cuff, Michael F. Lin, et al. "Distinguishing Protein-Coding and Noncoding Genes in the Human Genome." *Proceedings of the National Academy of Sciences* 104 (2007): 19,428–19,433. The researchers estimate there are 20,500 genes in the human genome.

Clark, David P., and Lonnie D. Russell. *Molecular Biology Made Simple and Fun.* 3d ed. St. Louis, Mo.: Cache River Press, 2005. This entertaining book explains the principles of molecular biology, which includes techniques and experiments with DNA, RNA, and proteins. The authors also discuss how these principles are applied in medicine or forensics.

ENCODE Project Consortium. "Identification and Analysis of Functional Elements in 1% of the Human Genome by the ENCODE Pilot Project." *Nature* 447 (14 June 2007): 799–816. The ENCODE Project Consortium describes its extensive study of a small but significant percentage of the human genome.

Fisher, Jasmin, Nir Piterman, Alex Hajnal, and Thomas A. Henzinger. "Predictive Modeling of Signaling Crosstalk during *C. elegans* Vulval Development." *Public Library of Science Computational Biology* (May 2007) Available online. URL: www.ploscompbiol.org/article/info:doi/10.1371/journal.pcbi.0030092. Accessed June 5, 2009. The researchers determined which genes and gene interactions are important if cells in *C. elegans* are to develop correctly.

Futreal, P. A., L. Coin, M. Marshall, T. Down, T. Hubbard, R. Wooster, et al. "A Census of Human Cancer Genes." *Nature Reviews Cancer* 4 (2004): 177–183. The authors list genes involved in cancer.

Gonick, Larry, and Mark Wheelis. *The Cartoon Guide to Genetics.* New York: HarperCollins, 1991. The Cartoon Guides comprise a series of books describing complex topics with simple figures and illustrations—cartoons. But these books are not cartoonish or childish, and the book on genetics in this series offers an accurate and entertaining overview of the subject.

Heller, Arnie. "Mining Genomes." Available online. URL: www.llnl.gov/str/June05/Ovcharenko.html. Accessed June 5, 2009. This article appeared in the June 2005 issue *Science & Technology,* a magazine published by Lawrence Livermore National Laboratory, in Livermore, California. The article offers information on the basics of bioinformatics, and also discusses bioinformatics research at Lawrence Livermore National Laboratory.

Kanehisa, Minoru, and Peer Bork. "Bioinformatics in the Post-Sequence Era." *Nature Genetics* 33 (2003): 305–310. The researchers review the present state of bioinformatics.

Korbel, Jan O., Alexander Eckehart Urban, Jason P. Affourtit, Brian Godwin, Fabian Grubert, Jan Fredrik Simons, et al. "Paired-End Mapping Reveals Extensive Structural Variation in the Human Genome." *Science* 318 (19 October 2007): 420–426. The researchers uncovered evidence of genetic variations with sizes larger than single nucleotides.

Maser, Richard S., Bhudipa Choudhury, Peter J. Campbell, Bin Feng, Kwok-Kin Wong, Alexei Protopopov, et al. "Chromosomally Unstable Mouse Tumours Have Genetic Alterations Similar to Diverse Human Cancers." *Nature* 447 (21 July 2007): 966–971. The researchers examined the DNA of the mice knock-outs and discovered genome instabilities with a striking similarity to those in cells of certain human tumors.

Moody, Glyn. *Digital Code of Life: How Bioinformatics Is Revolutionizing Science, Medicine, and Business.* New York: Wiley, 2004. Focusing on the business aspects of bioinformatics, this book explores how and why people and large corporations are heavily investing in this field of research.

Mouse Genome Sequencing Consortium. "Initial Sequencing and Comparative Analysis of the Mouse Genome." *Nature* 420 (5 December 2002): 520–562. This article reports the sequence of the mouse genome.

Pollard, Katherine S., Sofie R. Salama, Bryan King, Andrew D. Kern, Tim Dreszer, Sol Katzman, et al. "Forces Shaping the Fastest Evolving Regions in the Human Genome." *Public Library of Science Genetics* (October 2006). Available online. URL: www.plosgenetics.org/article/info:doi/10.1371/journal.pgen.0020168. Accessed June 5, 2009. The researchers found evidence that much of the genetic distinction between humans and chimpanzees does not lie in the coding part of genes but instead lies in their regulation.

Richards, Julia E., and R. Scott Hawley. *The Human Genome: A User's Guide.* 2d ed. Burlington, Mass.: Elsevier Academic Press, 2005. This books offers a thorough discussion of genetics and genomes, including heredity, molecular biology, mutations, and the Human Genome Project.

Watson, J. D., and F. H. C. Crick. "A Structure for Deoxyribose Nucleic Acid." *Nature* 171 (25 April 1953): 737–738. Available online. URL: www.nature.com/nature/dna50/watsoncrick.pdf. Accessed June 5, 2009. A classic paper that describes the double helix.

Web Sites

1000 Genomes. Available online. URL: www.1000genomes.org. Accessed June 5, 2009. Sponsored by an international consortium, the 1000 Genomes Project aims to sequence the genome of 1,000 people in order to study genetic variation in humans. A copy of the press release is on the Web site, as well as additional information about this ambitious project.

National Center for Biotechnology Information. Available online. URL: www.ncbi.nlm.nih.gov. Accessed June 5, 2009. A division of the National Library of Medicine, the NCBI maintains databases containing the human genome and other important biological data. The Web site of the NCBI contains links to the databases along with information on the NCBI's mission and its research.

———. "A Science Primer." Available online. URL: www.ncbi.nlm.nih. gov/About/primer/bioinformatics.html. Accessed June 5, 2009. This educational resource offers additional information on many of the subjects discussed in this chapter, as well as tutorials on a larger range of topics pertaining to genome and genetic research. Subjects include bioinformatics, genome mapping, molecular modeling, SNPs, microarrays, molecular genetics, and pharmacogenomics (the study of how a person's genome affects his or her response to drugs and medications).

National Human Genome Research Institute. Available online. URL: www.genome.gov. Accessed June 5, 2009. A member of the National Institutes of Health, the National Human Genome Research Institute funds and supports biomedical research on all aspects of the human genome, including bioinformatics research. Its home page contains much information on research, health issues, policies, and ethics, and also provides educational materials for students.

United States Department of Energy. Human Genome Project Information. Available online. URL: www.ornl.gov/sci/techresources/ Human_Genome/home.shtml. Accessed June 5, 2009. Visitors to this Web site will find links to a rich variety of resources on the Human Genome Project, including pages describing the goals and history of the project, research, medicine, genetics, ethics, legal issues, and social issues.

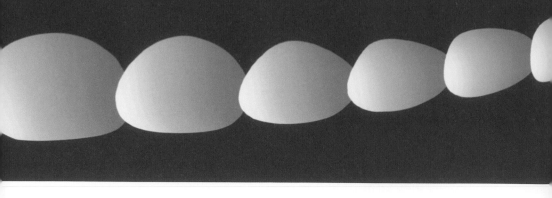

FINAL THOUGHTS

Computers have saved people countless time and effort by performing repetitious tasks involved in bookkeeping, word processing, data storage and retrieval, and much else. Advanced computers and algorithms have also begun to engage in more sophisticated tasks, as the chapters of this volume indicate. But a few jobs are so complex that they would seem to be forever in the exclusive domain of human beings. Perhaps foremost among those complex tasks is scientific discovery.

Science has come a long way in the past few centuries. Sir Isaac Newton (1642–1727) laid the foundation for much of physics, including essential mathematical techniques, in the 17th century. Chemistry made rapid strides in the 18th and 19th centuries, and in the 1860s Dmitry Mendeleyev (1834–1907) formulated the periodic table of chemical elements. Biological knowledge expanded dramatically in the 20th century when Francis Crick (1916–2004) and his colleagues discovered the genetic code. All the sciences continue to flourish in the 21st century as researchers from all over the world pursue a huge number of projects.

Achieving these scientific advances required a great deal of ingenuity, creativity, and hard work. Computers are certainly capable of hard work, but what about the ingenuity and creativity needed to make a new scientific discovery? This talent seems to be rare in people, and, just as importantly, it is not well understood. Scientists are not yet able to explain fully how the mind works even at the simplest level, much less do they understand creativity and ingenuity. This lack of understanding makes it extremely difficult to program a computer to do the job—no one knows exactly how the instructions should be written.

But the frontiers of computer science are continuing to break down barriers. Computers are gaining in speed and processing capacity, working more efficiently, and finding patterns and structure in data, especially in bioinformatics, where scientists are using computers to mine the huge databases accumulated in genomic research. Most relevant to scientific discovery, perhaps, is the implementation of artificial neural networks that can mimic the human brain and can even learn new things, including things that the engineer who designed the network did not know.

One of the foundations of scientific research is the ability to solve problems when the answer is not immediately apparent. To accomplish this, scientists tend to adhere to a specific method that involves gathering data, making testable hypotheses, and testing these hypotheses with experiments or observations. Knowledge and discovery eventually emerge, although not without a lot of effort (and an occasional lucky break).

A humanlike robot developed at SARCOS Research Corporation in Utah watches a poker game. *[Peter Menzel/Photo Researchers, Inc.]*

Herbert Simon (1916–2001) was a researcher who believed that computer systems were capable of doing science. A strong proponent of research in artificial intelligence, Simon said in a 1994 interview in *Omni,* "In general, when we've found something to be explainable in natural terms, we've found ways of building theories for it and imitating it. The Egyptians didn't think building pyramids was such a great thing for humans to do that they shouldn't develop machines to help. Why should we think that thinking is such a great thing that machines shouldn't help us with it?"

In 1983, Simon, then at Carnegie Mellon University, along with colleagues Gary F. Bradshaw and Patrick W. Langley, published a paper in *Science* titled, "Studying Scientific Discovery by Computer Simulation." The goal of this research was to determine if a computer program can make a scientific discovery: "First, we specify the problem-solving processes we think are used in and required for discovery, defining them with such specificity that we can write a computer program to execute them. Second, we confront the program with discovery problems that scientists have encountered, and we observe by whether the program can make the discovery, starting from the same point the scientists did, using only processes that, from other psychological evidence, are known to be within human capabilities."

The researchers used a computer program called BACON that they had developed to simulate critical aspects of scientific discovery. (The program was named after Sir Francis Bacon [1561–1626], a British philosopher who expounded the scientific method in the 17th century.) Given clues similar to those available to Joseph Black (1728–99), a chemist, BACON can make a discovery similar to the one Black made: "When provided with data about temperatures before and after two substances are brought into contact, the program infers the concept of specific heat and arrives at Black's law of temperature equilibrium."

The study of BACON provided insights into the discovery process. According to Simon, in his 1994 interview, "This tells you that using a fairly simple set of rules of thumb allows you to replicate many first-rank discoveries in physics and chemistry."

But BACON enjoyed certain advantages, such as the hindsight of the programmers. A major difficulty in science is to winnow all of the many possibilities into a few that can be studied in detail. In terms of data, this means separating the crucial information from the reams of observations that are not relevant to the problem at hand. Many a scientist has gotten sidetracked, following a false trail that led to nowhere.

Ross D. King, a researcher at the University of Wales, and his colleagues decided to see if they could create a "robot scientist" that could actually make new discoveries. They chose the field of genomics, which is fertile ground, rich in data. In their report, "Functional Genomic Hypothesis Generation and Experimentation by a Robot Scientist," published in *Nature* in 2004, the authors wrote, "The question of whether it is possible to automate the scientific process is of both great theoretical interest and increasing practical importance because, in many scientific areas, data are being generated much faster than they can be effectively analysed."

To test the idea, the researchers built an automated system that could handle biological material. Artificial intelligence software controls and operates the system. "The key point," wrote the researchers, "is that there was no human intellectual input in the design of experiments or the interpretation of data." The computer system does all the work: "The system automatically originates hypotheses to explain observations, devises experiments to test these hypotheses, physically runs the experiments using a laboratory robot, interprets the results to falsify hypotheses inconsistent with the data, and then repeats the cycle." In certain instances, such as the process of determining the function of genes in specific cells, the system "is competitive with human performance."

Scientists are still a long way from building machines to take their place. Most scientific problems call for a degree of insight and inventiveness well beyond the capacity of today's computer technology. But as the frontiers of computer science advance, the possibility of creating computer scientists—scientists that are actually computers—becomes more likely.

GLOSSARY

abacus   a calculating device consisting of rows of beads strung on wire and mounted in a frame

AI   *See* artificial intelligence

algorithm   step-by-step instructions to accomplish a certain task

alleles   alternate versions of a gene

artificial intelligence   the study or application of machines capable of creativity or making decisions and judgments similar to those a person would make in the same circumstances

axon   a long, thin projection from a neuron, along which signals travel

backpropagation   a method of training artificial neural networks in which the correction of response errors propagates backward through the network, from output to input

bases in DNA   and bioinformatics research, another term for *nucleotide*

binary   having two values, which in computer science are generally referred to as 1 and 0

bioinformatics   the use of computing technology to manage, search, and analyze biological systems and data

bit   binary digit, a unit of information with a value of either 1 or 0

byte   eight bits

chromosomes long, threadlike structures of DNA in the nucleus of a cell

cipher a form of encryption in which the meaning of the message is hidden by replacing each letter with another letter according to a specified scheme

code a form of encryption in which the meaning of the message is hidden by replacing each word or sentence with another word or some kind of symbol

codon three consecutive DNA bases that code for a specific amino acid

conductors materials such as metal that easily carry an electrical current

cryptanalysis the process of breaking codes or ciphers, or in other words, recovering the meaning of the message without having the key

cryptography the protection of a message to make it unintelligible to anyone not authorized to receive it—the sender renders it into an unintelligible form by processing it (this processing is known as **encryption**)

cryptology the science of writing, reading, and analyzing secret messages

data information, either in the form of facts and figures or instructions, such as computer programs, to tell a computer what to do; plural of *datum*

database a set of records that is stored and organized in ways to facilitate tasks such as searching for a specific item

decryption deciphering or decoding a message; turning it back into its original form

deoxyribonucleic acid molecules composed of units called bases that contain hereditary information

digital image data in the form of an array of picture elements called pixels, whose values represent the brightness or color at different points in the picture

DNA   *See* **deoxyribonucleic acid**

encryption   hiding the meaning of the message by using a code or cipher

entanglement   in quantum mechanics, the linking of the states of particles, so that measurement of the state of one particle allows the prediction of the state of another particle

exponential   raising a number to a variable power, such as 2^n, the value of which shows a huge increase as n gets bigger

factorization   decomposing a number into its factors, the product of which equals the number

floating point operations per second   a measure of a computer's speed in terms of the number of mathematical operations it can perform in a second

FLOPS   *See* **floating point operations per second**

frequency analysis   in cryptanalysis, the study of the frequency of letters in an attempt to decipher a secret message

genes   units of hereditary information consisting of specific sequences of DNA, which serves as instructions to synthesize important molecules

genome   the genetic material of an organism

heuristics   methods to help find the solution of a problem

IC   *See* **integrated circuit**

integrated circuit   an electrical component containing a large number of elements such as transistors

JPEG   a compression standard common format for image files that is used for photographic images. Files in this format often have the extension "jpg"

mainframe   a large general-purpose computer system, as opposed to a smaller minicomputer or microcomputer, used mainly by organizations for data processing, enterprise resource planning, and financial transaction processing

mRNA messenger RNA, a molecule of RNA that carries genetic information taken from DNA to the enzymes that make proteins

nanotechnology manipulating matter or building tiny machines on a very small scale

neural networks groups of neurons that process information

neuron a brain cell involved in information processing for sensation, perception, movement, and thinking

nucleotides units of DNA, which can be adenine (A), thymine (T), cytosine (C), and guanine (G)

phishing an attempt to obtain personal information from Internet users through fake Web sites or fraudulent e-mails

pixels picture elements, the ensemble of which compose the image

polynomial raising a variable to a fixed power, such as x^2, which increases as x gets bigger, but not as quickly as an exponential expression

postsynaptic neuron in a synapse, the neuron receiving the message

presynaptic neuron in a synapse, the neuron sending the message

protein a molecule composed of a sequence of units known as amino acids, which may serve for any of a variety of critical functions in organisms

quantum computers hypothetical devices that employ quantum mechanics to carry out calculations and other operations

quantum mechanics set of principles and theories in advanced physics that govern the behavior of small particles

qubit a bit of information in a quantum computer; pronounced "cubit"

ribonucleic acid a type of molecule that is similar in composition to DNA but does not have the same structure, which performs a number of functions, most of which are related to the synthesis of proteins

RNA *See* **ribonucleic acid**

scalability the ability to handle large amounts of data

semiconductors materials that have variable abilities to carry electrical currents, and are often used in the construction of devices such as computers that require manipulation of electrical signals

silicon an element widely used in semiconductors and computer technology

single nucleotide polymorphism a variation in a single DNA base

SNP *See* **single nucleotide polymorphism**

software programs that provide the instructions necessary for a computer to perform its functions

supercomputer a computer capable of working at a great speed that can process a very large amount of data within an acceptable time

superposition in quantum mechanics, the property of a system or a particle to be in a combination of states at the same time

synapses junctions between neurons through which the cells communicate with one another

synaptic weight the impact or influence a presynaptic neuron has on its postsynaptic target, or, in other words, how much emphasis it gives its message

Turing test a test proposed by mathematician and computer scientist Alan Turing, in which a machine is deemed intelligent if it can engage in conversation indistinguishable from that of a person, in the opinion of a human judge

FURTHER RESOURCES

Print and Internet

Alderman, John. *Core Memory: A Visual Survey of Vintage Computers.* San Francisco: Chronicle Books, 2007. Plenty of photographs and brief descriptions highlight this story of the evolution of computer technology, including punch cards—which was the way computer users once fed data and programs into computers—and giant, room-spanning machines.

Carr, Nicholas. *The Big Switch: Rewiring the World, from Edison to Google.* New York: W.W. Norton, 2008. Computers have transformed almost every aspect of society. In this book, Carr argues that the future of computer technology will resemble the historical transformation of the electrical power industry—a switch from local devices to a reliance on gigantic networks. This would be an important and somewhat unsettling change because it means that the network owners would be making all the decisions rather than the users.

Davis, Martin. *The Universal Computer: The Road from Leibniz to Turing.* New York: W.W. Norton, 2000. Many of the ideas of computers and computational logic began long before the arrival of the electronic machines of today. Logician and computer scientist Martin Davis relates the history of these ideas, and tells the story of the people who struggled to conceive them.

Dewdney, A. K. *The New Turing Omnibus.* New York: Holt, 1993. This book contains 66 brief essays that cover most of the topics of interest in computer science. The author categorizes the essays into 11 areas of computer theory and practice, including coding and cryptology, artificial intelligence, theory of computation, analysis of algorithms, and applications.

Dreyfus, Hubert L. *What Computers Still Can't Do: A Critique of Artificial Reason.* Cambridge, Mass.: Massachusetts Institute of Technology Press, 1992. An update of a 1972 book, Dreyfus offers a skeptical review of artificial intelligence research, and why he believes that machines will never be able to achieve more than a rough and unimpressive imitation of human intelligence.

Gralla, Preston. *How the Internet Works.* 8th ed. Upper Saddle River, N.J.: Que Publishing, 2006. The Internet is one of the most prominent computing applications and can also be one of the most puzzling. This books helps clear up some of the mysteries.

Hall, J. Storrs. *Beyond AI: Creating the Conscience of the Machine.* Amherst, N.Y.: Prometheus Books, 2007. As computer scientists and artificial intelligence researchers create increasingly sophisticated machines, people have started to wonder how far these machines can go, and what their relation with human society will be in the future. Hall, a computer scientist, provides an overview of the subject.

Ifrah, Georges. *The Universal History of Computing: From the Abacus to the Quantum Computer.* New York: Wiley, 2002. This book covers a broad swath of history, from the earliest concepts to the most recent advances. The author focuses on numbers, computing, and computational methods rather than machines.

Lévénez, Éric. "Computer Languages History." Available online. URL: www.levenez.com/lang. Accessed June 5, 2009. Computer scientists have developed many languages with which to write programs. This compilation provides a timeline as well as links that supply further information on languages such as C, Java, PASCAL, and many others.

Levinson, Paul. *The Soft Edge: A Natural History and Future of the Information Revolution.* London: Routledge, 1997. The ways in which people store, transmit, and manipulate information has changed over time, and inventions such as the computer, along with the telegraph, telephone, and many others, have had drastic effects. Levinson explains how these changes have wielded enormous influence on how people live and interact, and what kind of changes may be expected in the future.

McCarthy, John. "What Is Artificial Intelligence?" Available online. URL: www-formal.stanford.edu/jmc/whatisai/whatisai.html. Accessed June 5, 2009. McCarthy, a retired professor of computer science, has compiled a list of questions and answers covering all aspects of artificial intelligence.

Muuss, Mike. "Historic Computer Images." Available online. URL: ftp.arl.mil/ftp/historic-computers. Accessed June 5, 2009. Computers have changed dramatically in size and appearance since the 1940s. This Web site contains a collection of high-quality digital images of early computers.

Petzold, Charles. *The Annotated Turing: A Guided Tour through Alan Turing's Historic Paper on Computability and the Turing Machine.* Indianapolis: Wiley Publishing, 2008. Turing's ideas had a tremendous impact on the development of computers and computer science, but they are not easy to follow. This book provides some much needed help.

Time. "The *Time* 100—Alan Turing." Available online. URL: www.time.com/time/time100/scientist/profile/turing.html. Accessed June 5, 2009. An entry in the "*Time* 100—The Most Important People of the Century," this brief biography of Alan Turing discusses the life and work of this computer pioneer.

Web Sites

History of Computing Foundation. The History of Computing Project. Available online. URL: www.thocp.net. Accessed June 5, 2009. This Web site contains an illustrated time line of computer development and an impressive collection of biographies of important computer scientists.

How Stuff Works. Available online. URL: www.howstuffworks.com. Accessed June 5, 2009. This Web site hosts a huge number of articles on all aspects of technology and science, including computers.

San Diego Supercomputer Center. Available online. URL: www.sdsc.edu. Accessed June 5, 2009. Founded in 1985, the San Diego Supercomputer Center is a unit of the University of California at San Diego, and provides high-speed computing resources to thousands of researchers in many different fields of study. The center's home page offers information on the computing resources as well as the research results and discoveries that the center has facilitated.

ScienceDaily. Available online. URL: www.sciencedaily.com. Accessed June 5, 2009. An excellent source for the latest research news, ScienceDaily posts hundreds of articles on all aspects of science. The articles are usually taken from press issues released by the researcher's institution or by the journal that published the research. Main categories include Computers & Math, Health & Medicine, Mind & Brain, Matter & Energy, and others.

University at Albany, State University of New York. Computer Science. Available online. URL: library.albany.edu/subject/csci.htm. Accessed June 5, 2009. This Web site contains an excellent collection of links to computer science topics, including programming and programming languages, algorithm collections, history, software, and many more.

University of Pennsylvania. The ENIAC Museum Online. Available online. URL: www.seas.upenn.edu/~museum. Accessed June 5, 2009. The Electronic Numerical Integrator and Computer (ENIAC), finished in 1946 and installed at the University of Pennsylvania, was one of the earliest modern computers. The University's School of Engineering and Applied Science maintains a museum devoted to ENIAC, along with a Web site that describes the machine and how it was built.

INDEX